# Forensic Applications of

# High
# Performance
# Liquid
# Chromatography

# Forensic Applications of
# High Performance Liquid Chromatography

Shirley Bayne
Michelle Carlin

CRC Press
Taylor & Francis Group
Boca Raton   London   New York

CRC Press is an imprint of the
Taylor & Francis Group, an **informa** business

CRC Press
Taylor & Francis Group
6000 Broken Sound Parkway NW, Suite 300
Boca Raton, FL 33487-2742

© 2010 by Taylor and Francis Group, LLC
CRC Press is an imprint of Taylor & Francis Group, an Informa business

No claim to original U.S. Government works

Printed in the United States of America on acid-free paper
10 9 8 7 6 5 4 3 2 1

International Standard Book Number: 978-1-4200-9191-5 (Paperback)

**Library of Congress Cataloging-in-Publication Data**

Bayne, Shirley.
    Forensic applications of high performance liquid chromatography / Shirley Bayne, Michelle Carlin.
        p. cm.
    Includes bibliographical references and index.
    ISBN 978-1-4200-9191-5 (hardcover : alk. paper)
    1. Liquid chromatography. 2. Chemistry, Analytic. 3. Forensic sciences. I. Carlin, Michelle. II. Title.

QD79.C454B39 2010
363.25'62--dc22                                                          2009043223

Visit the Taylor & Francis Web site at
http://www.taylorandfrancis.com

and the CRC Press Web site at
http://www.crcpress.com

# Table of Contents

## 11    Forensic Applications of HPLC                     213

# Preface

This book has been written for university students studying analytical chemistry, applied chemistry, forensic chemistry, or other such courses where there is an element of HPLC within the course curriculum. The aim of the book is to explain HPLC from a forensic science perspective, and many of the examples used here are associated with real-life samples that might be expected within a forensic science laboratory. We have tried to maintain a balance between practical solutions and the theoretical considerations involved in HPLC analysis. The book takes the reader on a journey through the world of HPLC; it is suitable for first-time users as well as those pursuing postgraduate study or in the early stages of their forensic analysis careers.

Many of the applications within forensic science adopt a reverse phase (RP) HPLC mode of separation in relation to analyses. We have chosen to use example applications of HPLC based on this particular mode of chromatography because it is the most frequently used. We have broken each chapter down into theoretical considerations with examples where appropriate, a key point summary, a series of questions where appropriate, and, finally, a list of books and journal articles that we believe will give further, thorough insight into each of the topics covered. We have attempted to keep the book as succinct as possible whilst still providing sufficient information to allow the reader to develop his or her knowledge at an effective pace.

In Chapter 1, we have included a brief history of HPLC because we believe that it is important to understand just how far the technique has advanced in its relatively short lifetime. We hope that this will inspire others to be innovative and explore a continually advancing field of study with huge opportunities. We have provided but a snapshot in time.

In Chapter 2, we move on to the theory behind the separation process. Before any judgement on an analysis can be made, it is necessary first to understand a little bit about the processes taking place. It is only by doing this that one can begin to understand when things have gone wrong and can then put them right. Forensic science requires a rigorous approach to the analysis of evidence, and the consequences of an error are far reaching. The basic chromatographic processes are discussed along with the chemistry theory that underpins this.

In Chapter 3, we concentrate on some of the basic requirements that will help to ensure a successful analysis, such as sample preparation and mobile phase preparation. We explore some of the limitations associated with these practical elements of HPLC analysis and provide information relating to current best practice.

In Chapter 4, we look at the different modes of separation that can be used in forensic science. We acknowledge that other separation chemistries do exist; however, these are beyond the scope of this primer. We have chosen to focus on reverse phase, normal phase, and ion exchange chromatography because these cover most of the mainstream applications.

In Chapter 5, we look at modes of detection, again examining those that we feel are best suited to forensic applications.

In Chapter 6, we move into the world of HPLC method development and have chosen to focus on RP-HPLC for the reasons given earlier. This chapter allows us to apply the theory from preceding chapters to more practical elements of HPLC analysis. It also allows us to explore in greater detail some of the many factors that need to be considered and the consequences of the different aspects within this field of study.

In Chapters 7 and 8, we look at how we ensure that our methodology is going to give us a true representation of what is present in the samples that we are analysing. This requires that the operator have knowledge of the other systems that are in place to support the HPLC analysis, such as the processes of validation, qualification, and estimations of error.

Chapter 9 covers the quality aspects of laboratory operation in general, and much of this theory can be applied to any analytical method. Most laboratories will operate with at least some of the elements discussed in this chapter.

Chapter 10 deals with troubleshooting HPLC systems and analyses. It is of fundamental importance that any analyst reporting results to the prosecution services in connection with an alleged offence be 100% sure of the validity of his or her analytical data. Ensuring that this happens means knowing when something is not right and being able to put it right. In this chapter, we highlight a series of common problems that can be encountered in HPLC and provide a number of possible solutions at each stage.

Chapter 11 looks at some of the applications of HPLC within the field of forensic science in greater detail. We have covered the most common areas, but we acknowledge that both HPLC and forensic science are far-reaching fields of study.

We hope that the book remains a companion throughout the reader's studies and we wish our readers well in their careers.

We would like to thank our families and friends, who have provided vision and commitment and been supportive throughout the writing process. We would also like to thank the following people for their valuable time and expertise during the review process; their insightful comments were much appreciated: Dr Joseph McGinnis, Mrs Helen Hodgson, Mr Douglas McLellan, Mr Ed Ludkin, and Mr Stephen Sole. Our project student, Mr Olivier Weiss, is thanked for his valuable contribution. Lastly, we would like to give special thanks to Dr Liam O'Hare for his contribution, skills, and knowledge.

**Shirley Bayne**
**Michelle Carlin**

# The Authors

**Shirley Bayne** studied applied chemistry in Newcastle upon Tyne and graduated with an honours degree. She started her career with the Medicines Testing Laboratory (MTL; part of the Royal Pharmaceutical Society of Great Britain) in Edinburgh, where she was given the opportunity to use a variety of different analytical techniques. Her responsibility for aspects of HPLC systems within the lab sparked her interest in this technique in particular. On leaving MTL, Shirley joined Lothian and Border Police Forensic Science Laboratory as a reporting analyst in the chemistry and drugs department. After a short spell leading the chemistry team, Shirley changed police forces and continued her forensic science career.

In 1998, Shirley joined the contract research company Quintiles Ltd. in its Pharmaceutical Method Development Department. She supervised a team responsible for the development and validation of HPLC assays for new drug entities. A number of years later, Shirley moved on to join BioReliance Ltd as analytical services manager. Part of her role involved leading a team of analysts responsible for the method development, validation, and transfer of release testing methods for biopharmaceutical products. In 2005, Shirley moved into academia at Teesside University as a senior lecturer and course leader in forensic science.

**Michelle Carlin** studied at Heriot-Watt University on the honours programme in colour chemistry, with a spell in a dyehouse in the Scottish Borders before embarking on a career in analytical chemistry. After some time spent in a contract research organisation in Edinburgh, Michelle went on to continue her education with an MSc in forensic science from Strathclyde University. She carried out a research project in the toxicology department of the Institut de Recherche Criminelle de la Gendarmerie Nationale (IRCGN) in Paris, using LC–ESI–MS.

After this, Michelle became the manager of a workplace drug testing laboratory in northeastern England before taking up a teaching position as lecturer in forensic science at Teesside University for 3 years. In 2009, she moved to Northumbria University as a junior lecturer in forensic chemistry, where she carries out research in analytical toxicology.

# Introduction to HPLC

1

## Introduction

High performance liquid chromatography is the topic of this primer, but the starting point for this goes much further back than modern times. Chromatography is a technique that has a number of forms, such as thin layer chromatography (TLC), high performance liquid chromatography (HPLC), and gas liquid chromatography (GLC, although more commonly known as GC). Each of these forms of chromatography has a variety of uses in the analytical sciences.

In forensic science, chromatography is used in the analysis of drugs of abuse, toxicology, fire debris analysis, environmental analysis, and explosives analysis, to name but a few. To understand each of the chromatographic techniques, especially HPLC as the topic of this primer, it is necessary first to explain what chromatography is and the basic principles of chromatography.

## The History of Chromatography

Chromatography seems to have been around for a long time in its current state, so it is difficult to believe that the first reported work naming and using chromatography was just over 100 years ago. Mikhail Semenovich Tswett (1872–1919) was born in a small town in Italy but grew up in Switzerland with his father. In later life, he went on to study botany at university in Geneva and, by 1896, he had completed the work and write-up for his doctoral thesis, for which he carried out work on the structure of the plant cell, chloroplasts, and the movement of protoplasm.

In 1896, he moved to Russia to join his father, but Tswett had difficulty finding the academic position for which he had wished. He accepted a temporary position in a laboratory and completed work for a magisters degree in order to gain the qualification to apply for academic positions in Russia. During this research, he started building the foundations to develop chromatography as a technique.

Tswett's research involved the isolation of chlorophyll from plant material. Whilst carrying out his work, he found a difference in polar and nonpolar solvents as to how well the isolation occurred. He concluded that this phenomenon was not due solely to the solubility of the chlorophyll in the solvent chosen. Rather, the separation was also affected by the interaction of the chlorophyll with the molecular forces at work in the plant cells.

From this point, Tswett went on to investigate more than 100 materials of organic and inorganic natures that might have been used as adsorbents. He presented his work in 1903, but he did not provide full details until he completed further work, which resulted in two fundamental papers in 1906 in the *Bulletin of the German Botanical Society*. This was the first time that 'chromatography' was mentioned.

Sadly, the work carried out by Tswett was not accepted by his peers. However, within 10 years of his death, the significance of the technique was realised and many applications were found across the sciences. Although some scientists ridiculed the work of Tswett, others used his findings to apply chromatography in their work (e.g., Leroy Sheldon-Palmer, Archer Martin, and Richard Synge).

Leroy Sheldon-Palmer (1887–1944) investigated carotenoids in butter; this was one of the first applications of Tswett's technique. In 1922, Sheldon-Palmer published a book on his work with carotenoids that brought chromatography to the attention of scientists outside Russia. This subsequently led to the reinvestigation of the technique in the 1930s.

Shortly after, the most prominent work was completed by Archer Martin (1910–2002) and Richard Synge (1914–1994) from Cambridge University. These scientists used silica gel and water in a glass column as a stationary phase and used chloroform as a mobile phase for the separation of amino acid derivatives. This was the beginning of *partition chromatography*. Martin and Synge were awarded the Nobel Prize in 1952 for their contribution to the field. Their work led to the development of other chromatographic methods, such as paper chromatography (1944), thin-layer chromatography (1956), and, eventually, high performance liquid chromatography (1967–1969).

## Basic Instrumentation

The form that liquid chromatography takes today in the twenty-first century would be unrecognisable to Tswett. Yet, the basic principles of the technique identified in his work in the early twentieth century still stand (Figure 1.1). It is no longer necessary to pack glass columns manually in order to carry out a separation. Nowadays, we have very sophisticated commercially packed columns and instruments to control the separations. These commercial HPLC instruments are composed of a number of modules that will be used to effect a separation, detect the compounds, and display the resulting chromatograms. Each of these modules will be discussed in subsequent chapters of this book, but we shall provide an overview of the HPLC before continuing with discussions on the separate components:

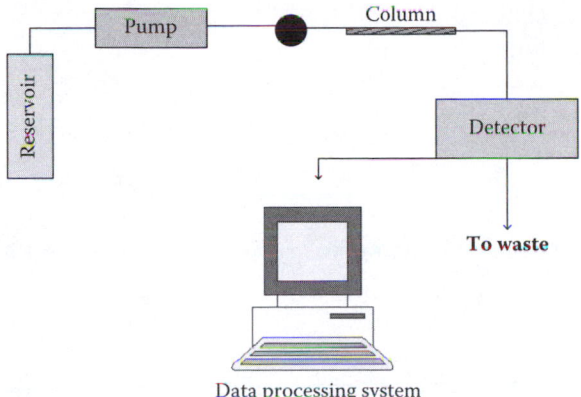

**Figure 1.1** (**See colour insert following page 142.**) Block diagram of HPLC instrument.

The *solvent reservoir* contains the mobile phase, which is typically made up of mixtures containing water and an organic modifier such as acetonitrile or methanol and/or buffer solutions. However, complex separations may require more complex mobile phase compositions. The mobile phase assists in the transportation of compounds in a mixture through the stationary phase, effecting separation into the individual components. This will be further discussed in Chapters 3 and 4.

The *pump* forces the mobile phase through the column at high pressure and is monitored and controlled by the pressure regulator. Different pump types are available; binary and quaternary pumps are the most frequently employed. A binary pump can drive two mobile phases at the same time, whereas a quaternary pump can drive four different mobile phases at the same time, if required. The pump must be able to generate high pressures, deliver flow rates from 0.1 to 10 mL min$^{-1}$, have reproducibility of 0.5% (or better), and be resistant to corrosion by a variety of solvents.

The *injector port* allows controlled introduction of the liquid sample into the HPLC instrument via a rotating valve system.

The *column* contains the packed stationary material and works with the mobile phase to effect the separation of the mixture into its individual components. Typically, the column operates at room temperature; however, it can be heated or cooled depending on the application. For changes in column temperature, however, an HPLC instrument with a column heating/cooling block must be purchased. Columns vary in length, packing particle size, and internal diameter. Some examples of column lengths are shown in Figure 1.2.

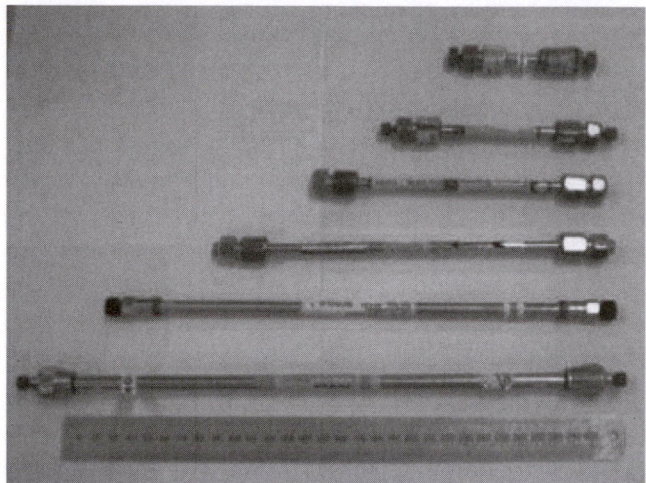

**Figure 1.2** Different lengths of a chromatography column.

The *detector* type will vary depending on the application; however, the most commonly used systems are diode array UV/Vis. Other detectors are available, such as electrochemical and mass spectrometry detectors. Detectors used will vary depending on the type of chromatography and the type of analyte required to be detected (e.g., when carrying out ion exchange chromatography, an electrochemical detector is required because a UV/Vis detector will not detect the ions). This will be discussed further in Chapter 5.

The *data processing system* allows for the manipulation of the signal response from the detector into a readable format in the form of a chromatogram. Software will be included from the company from which the HPLC instrument was purchased. There are differences from manufacturer to manufacturer; however, all software will control the setup of the instrument and will allow manipulation of data when the analyses are complete.

## Further Reading

Isaaq, H. J. (ed.). 2001. *A century of separation science.* Boca Raton, FL: CRC Press.
Livengood, J. 2009. Why was M. S. Tswett's chromatographic adsorption analysis rejected? *Studies in History and Philosophy of Science* 40: 57–69.
The centenary of chromatography. *Journal of Analytical Chemistry* 58 (8): 703–705 (translated from *Zhurnal Analiticheskoi Khimii* 58 (8), 2003, 789–791).

# Basic Principles of HPLC

2

## Theory of Chromatography

High performance liquid chromatography (HPLC) takes many different forms and is a primary method of analysis in many types of laboratories. HPLC separates mixtures of compounds into their individual components by means of an interaction of the compound, a liquid mobile phase, and an inert stationary phase. The interactions between the mobile phase and stationary phase are diverse and dependent on the types of compounds undergoing separation and the extent to which separation is required. The output from the data analysis system is a *chromatogram*. An example of this is shown in Figure 2.1, where the small peak at the beginning is the peak for the unrestrained solute and shows the point at which the injection has been detected (change in pressure of the system). The x-axis represents the time and the y-axis represents the detector response.

A number of modes of separation are used in liquid chromatography. The commonly used modes in forensic science are reversed phase (RP), normal phase (NP), and ion exchange (IEC). Each of these will be discussed in greater detail in Chapter 4.

Separation of compounds takes place in the HPLC stationary phase packed within the column (usually stainless steel, but other housing materials

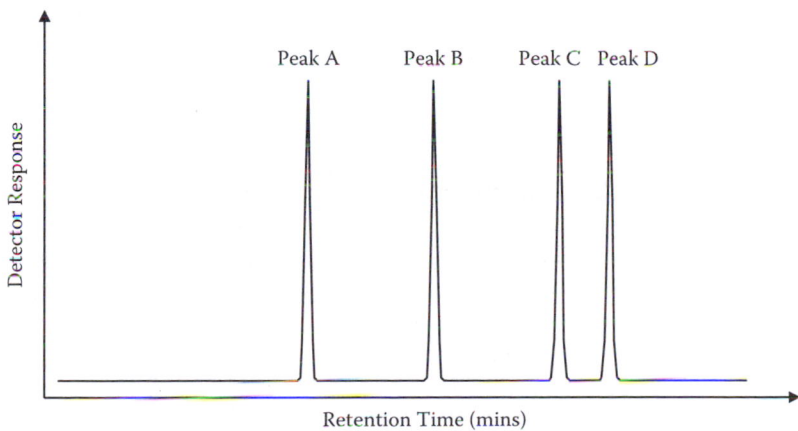

**Figure 2.1** Basic chromatogram showing a four-component mixture.

5

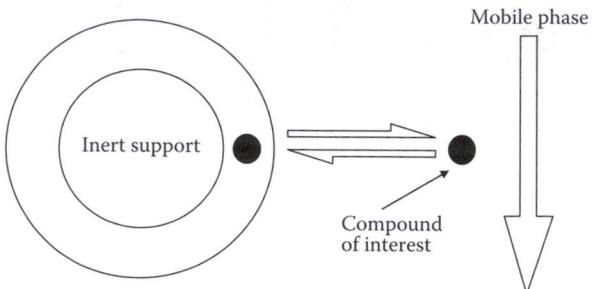

**Figure 2.2** $C_{stat}$, $C_{mob}$.

are available). The time that it takes for a compound to pass through the column and reach the detector is called the *retention time* ($t_R$). Ideally, each compound in the sample mixture will have a different retention time.

The separation is dependent on the distribution of the compounds between the stationary and the mobile phases and can be represented by the *distribution coefficient, K,*

$$K = \frac{C_{stat}}{C_{mob}}$$

where $C_{stat}$ is the concentration of compound X in the stationary phase and $C_{mob}$ is the concentration of compound X in the mobile phase (see Figure 2.2).

A compound that has a higher affinity for the stationary phase will take longer to travel through the column than one that has a higher affinity for the mobile phase. In order for separation to occur, the distribution coefficient for each of the compounds must be different for a given mobile and stationary phase. Some of the chemical properties of solutes that affect $K$ will be discussed later in this chapter.

### Retention Factor

The retention factor ($k'$), which was previously known as the capacity factor, represents the capacity of a stationary phase to attract an analyte. $k'$ can be defined as the time during which the sample remains in the stationary phase relative to the time during which it resides in the mobile phase ($t_M$), as shown in Figure 2.3. The retention factor is represented by the following equation:

$$k' = \frac{t_R - t_M}{t_M}$$

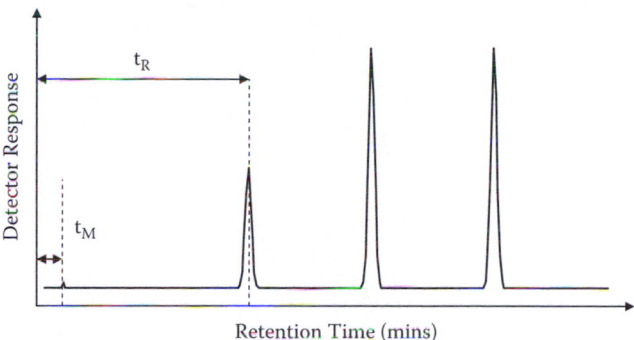

**Figure 2.3** Retention factor.

Let us consider the following hypothetical example.

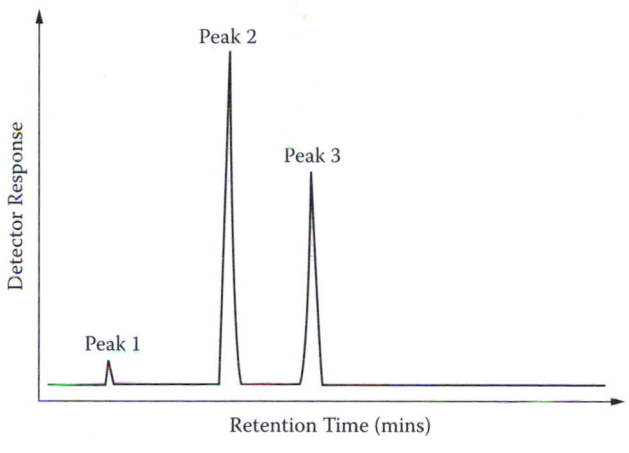

| Peak | Retention time (min) |
|---|---|
| 1 ($t_m$) | 1.1 |
| 2 | 3.3 |
| 3 | 4.6 |

What are the $k'$ values for the analytes eluted at 3.3 and 4.6 minutes? We can calculate these using the formula

$$k' = \frac{t_R - t_M}{t_M}$$

$k'$ for peaks 2 and 3 can be determined as follows:

Peak 2:

$$k' = \frac{3.3 - 1.1}{1.1}$$

$$k' = 2.7$$

Peak 3:

$$k' = \frac{4.6 - 1.1}{1.1}$$

$$k' = 3.2$$

Here, we see that the longer the analyte is retained (i.e., the longer the retention time), the greater is the value of $k'$. It is generally accepted amongst chromatographers that a $k'$ value of between 1 and 10 is an optimum value. A value of less than 1 means that the compound is effectively unretained or that no interaction is taking place with the stationary phase. This is an undesirable situation because no chromatography is taking place (i.e., nothing is being retained by the column). $k'$ can be used rather than retention time because it is more reproducible from run to run. This is due to the fact that $k'$ is independent of the velocity of the mobile phase and therefore less sensitive to fluctuations in the HPLC system.

---

**EQUATION WARNING**

The following section contains a lot of equations. If you are interested in the physical chemistry of chromatography, it will help you to understand how some important practical variables have an impact on the retention time of your analyte. Indeed, you may want to look for more detailed and rigorous treatments of these phenomena, such as those found in the 'Further Reading' section at the end of the chapter.

On the other hand, you may find this section to be too theoretical for your purposes. In that case, you can skip to the summary, in which the main practical outcomes of this chapter are described but not explained.

---

## Retention Time and Retention Volume

The retention time ($t_R$) has already been defined; it is the time taken for a solute to migrate from the injector, or point of introduction, to the detector. At a constant flow rate, this is simply related to the retention volume ($V_R$), which is the volume of mobile phase that passes through the column in this time; thus,

$V_R = t_R F$, where $F$ is the flow rate, usually in millilitres per minute. For an unretained solute, the retention volume is known as the void volume ($V_o$ or $V_M$).

The retention of solutes has also been considered in terms of their distribution coefficient ($K$), which is the ratio of the *concentrations* of the analyte in the mobile phase ($C_{mob}$) and in the stationary phase ($C_{stat}$) at equilibrium. In this section, we will also consider the retention factor or capacity factor. Like the distribution coefficient, $k'$ is dependent on the relative affinity of the solute in the respective phases, but it is expressed as the ratio of the *amount* of solute in stationary and mobile phases. Thus,

$$k' = \frac{n_{stat}}{n_{mob}}$$

where $n_{stat}$ and $n_{mob}$ are the quantities of solute present in the stationary and mobile phases, respectively.

Assuming that $n_{stat}$ is the concentration of solute in the stationary phase ($C_{stat}$) times the volume of stationary phase ($V_{stat}$) and $n_{mob}$ is similarly the concentration of solute in the stationary phase ($C_{mob}$) times the volume of stationary phase ($V_{mob}$), then

$$k' = \frac{C_{stat} V_{stat}}{C_{mob} V_{mob}}$$

Because

$$K = \frac{C_{stat}}{C_{mob}}$$

then

$$k' = K \frac{V_{stat}}{V_{mob}}$$

For any particular column, $V_{stat}$ and $V_{mob}$ are fixed values (at least for the duration of a particular analysis), so $k'$ is directly related to $K$ (distribution coefficient). This makes it a useful parameter to measure, but in this form it is not any easier to measure than $K$ itself. In order to get a definition of $k'$ that allows us to make practical measurements, we need to consider $f$ as the fraction of time spent in the mobile phase. It follows (with a bit of thought) that $f$ is also the fraction of the solute present in the mobile phase at any one time. If the total amount of solute is $n_{stat} + n_{mob}$, then

$$f = \frac{C_{mob} V_{mob}}{C_{mob} V_{mob} + C_{stat} V_{stat}}$$

This can be rearranged, with the inclusion of $K$, to give

$$f = \frac{1}{1 + \dfrac{C_{stat} V_{stat}}{C_{mob} V_{mob}}}$$

$$f = \frac{1}{1 + K \dfrac{V_{stat}}{V_{mob}}}$$

$$f = \frac{1}{1 + k'}$$

Because $f$ determines the retention time of a given solute, $k'$ can now be directly related to retention time. As discussed previously, when the solute is associated with or dissolved in the stationary phase, its velocity is 0. When it is in the mobile phase, its velocity is equal to that of the mobile phase ($u$). Thus, the average velocity ($u_a$) depends only on the velocity of the mobile phase ($u$) and the proportion of time spent in the mobile phase ($f$):

$$u_a = uf$$

$$u_a = \left( \frac{1}{1 + k'} \right) u$$

$$u_a = \frac{u}{1 + k'}$$

Remembering again not to confuse the velocity of mobile phase (in centimetres per minute) with flow rate (in millilitres per minute), it is fairly obvious that the time required for a solute to travel down a column (the retention time, $t_R$) is the length of the column ($L$) divided by the average velocity of the solute ($u_a$):

$$t_R = \frac{L}{u_a}$$

$$t_R = \frac{L \left( 1 + k' \right)}{u}$$

However, because $L/u$ is obviously the retention time of an unretained solute, which we have previously defined as $t_M$,

$$t_r = t_M \left(1 + k'\right)$$

Rearranging this yields:

$$t_R = t_M + k' t_M$$

$$t_R = t_M = k' t_M$$

$$\frac{t_R = t_M}{t_M} = k'$$

Thus, if we know $t_R$ and can measure $t_M$ under the same conditions, it is easy to calculate the capacity factor ($k'$), which, as we found earlier, is directly related to the fundamental distribution coefficient ($K$) that determines the retention of the solute.

## Separation Factor

The separation factor ($\alpha$) is used to describe the position of two peaks relative to each other and is measured using the following equation:

$$\alpha = \frac{k'_2}{k'_1}$$

This factor is calculated using the retention factor ($k'$) for each of the component peaks, where

$$k' = \frac{t_R - t_M}{t_M}$$

The separation factor (Figure 2.4) is dependent on a number of factors, such as the nature of the stationary phase, the mobile phase, temperature, and the compounds of interest.

Using the preceding example, we can calculate $\alpha$ for peaks A and B, where $t_{RB} = 5.00$, $t_{RA} = 2.80$, and $t_M = 1.2$:

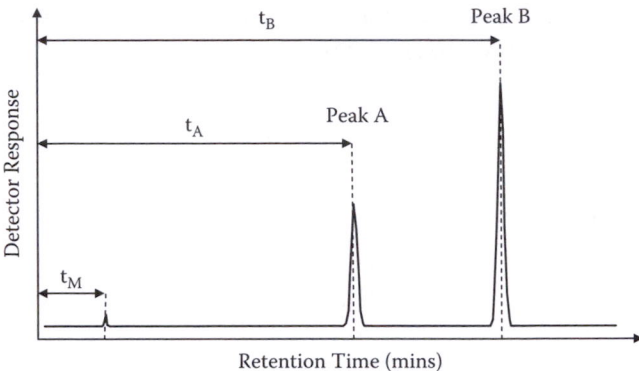

**Figure 2.4** Separation factor.

Peak B:

$$k'_B = \frac{5.00 - 1.2}{1.2} = 3.17$$

Peak A:

$$k'_A = \frac{2.80 - 1.2}{1.2} = 1.33$$

$$\alpha = \frac{3.17}{1.33} = 2.4$$

## The Effect of Temperature on Retention

As with any equilibrium constant, the effect of temperature on $K$ (distribution coefficient) is described by the van't Hoff equation:

$$\frac{d\ln K}{dT} = \frac{\Delta H}{RT^2}$$

where $\Delta H$ is the enthalpy of the reaction (in this case, enthalpy of solution, or the enthalpy of adsorption if that better describes the mechanism of retention), $R$ is the gas constant, and $T$ is the temperature in kelvins.

We have already assumed that $V_{stat}$ and $V_{mob}$ are constant and thus have seen that $K$ is proportional to $k'$. Extending this assumption, if $V_{stat}$ and $V_{mob}$ are also independent of temperature, then the rate of change of $K$ with temperature will be equal to the rate of change of $k'$. Therefore,

$$\frac{d\ln k'}{dT} = \frac{\Delta H}{RT^2}$$

Making the approximation that, over the fairly narrow range of temperatures at which HPLC is used $\Delta H$ will be independent of temperature, we can take the integral

$$\ln k' = \text{constant} - \frac{\Delta H}{RT}$$

This shows that $\ln k'$ is inversely proportional to temperature. This has the important implication that, at higher temperatures, retention times are shorter. In GC, this is used to control retention time, but in HPLC it is an often overlooked source of error. Modern systems often include a column heater that is intended to maintain a constant temperature, thus reducing this error. These are more or less essential if the intention is to run a system overnight using an autosampler because the usual decrease in laboratory temperature during the night can significantly alter retention times, causing methods that have been carefully set up to fail. Remember also that the mobile phase reservoir will also cool during the night; check that the column heater has a preheating section that warms the mobile phase to the column temperature before it enters the column.

## SOME PRACTICAL WARNINGS REGARDING VARIATIONS IN TEMPERATURE (INCLUDING USE OF COLUMN HEATERS)

Many methods now specify the use of a column heater to run separations at higher temperatures, thus decreasing retention times and obtaining more rapid results. This can be very effective, but it is important to remember the following:

- Some of the processes that give rise to band broadening are also affected by temperature. In particular, longitudinal diffusion will be more noticeable at higher temperatures. This is not usually a major problem and is normally outweighed by the advantages.
- If traditional silica-based HPLC columns are used, the solubility of the stationary phase will be increased at higher temperatures, and this can have a dramatic impact on column life. Because silica is more soluble at extremes of pH, this can be especially significant when considered along with the following point.
- pH measurements are temperature dependent and pH metres need to be carefully calibrated at the temperature at which they are to be used (follow the manufacturer's instructions carefully). Also, the $K_w$

of water changes with temperature, and the effect is to alter the pH at which a solution can be considered neutral, from 7.00 at 25°C to 6.77 at 40°C. Most common buffers are temperature dependent, so a buffer prepared to have a particular pH at room temperature may have a very different pH at a higher temperature. Thus, the mobile phase may not have the pH that it is thought to have as it travels down a heated column. Often the pH will be lower than expected. As well as significantly altering the retention times of some solutes, this can have an impact on the stationary phase. Similar considerations are important if chromatography is carried out at reduced temperatures (e.g., to preserve sensitive analytes or samples prone to biological degradation).

- Some mobile phases are prone to microbial contamination. Acetate buffers are particularly susceptible. This may not be a significant problem if the mobile phase is stored at a low temperature and then heated in a column heater; however, if the buffer is stored for a few hours at 25°C or above, it may be altered as a result of microbial growth.

## Reporting Retention Times

Retention times can be quoted in a number of ways (see Table 2.1):

Retention time ($t_R$). The simplest, and possibly the most commonly used, is simply the retention time ($t_R$) in minutes. However, $t_R$ values are very much dependent on local parameters, such as details of the particular column, temperature, and flow rate. This makes comparison of $t_R$ values obtained in different labs or even on different days in the same lab unreliable.

Elution volume ($V_R$). A more consistent measure is the retention volume or elution volume ($V_R$). This is better because it is independent of flow rate. Unfortunately, the situation where it is most useful to

**Table 2.1   Summary Effect of an Increase in Parameter on $t_R$**

| Symbol | Parameter | Change in Retention Time Due to Increase in Parameter |
|--------|-----------|-------------------------------------------------------|
| $T$ | Temperature | Decrease |
| $V_{mob}$ | Volume of mobile phase | Decrease |
| $V_{stat}$ | Volume of stationary phase (or surface area of stationary phase) | Increase |
| $K_D$ | Distribution coefficient (note that distribution coefficient depends on other factors) | Increase |
| $u$ | Solvent velocity | Decrease |
| $L$ | Column length | Increase |

measure $V_R$—when the flow rate is not well controlled or is deliberately varied—is the situation where $V_R$ is most difficult to calculate. $V_R$ is not much used to report HPLC data; because flow rates are usually reliable, it is more common to report the flow rate along with $t_R$, allowing $V_R$ to be calculated if necessary. $V_R$ values are commonly quoted in preparative chromatography, including preparative HPLC, or where fractions of eluate are collected for further analysis.

*Adjusted retention time* $(t_R')$. The adjusted retention time $(t_R')$ and the adjusted retention volume $(V_R')$ are sometimes quoted. They are calculated by subtracting the retention time or retention volume of an unretained solute $(t_M$ or $V_M)$:

$$t_r' = t_R - t_M$$

*The retention (capacity) factor.* If $t_M$ has been determined in order to calculate the adjusted retention time, it is more common to use it to calculate and quote the retention factor $(k')$ according to the following equation:

$$k' = \frac{t_R - t_M}{t_M}$$

This is one of the best values to quote for characterisation of retention behaviour because it is relatively independent of flow rate, temperature, and other variables. It is not entirely transferable between analyses because the distribution coefficient $(K)$ is normally slightly temperature dependent (despite the assumption made earlier). However, $K$ for an unretained solute is, by definition, 0 and will be 0 at any temperature; thus, the calculation of $k'$ does not take wide variations of temperature into account. Where the range of temperatures is narrow, as is usually the case for HPLC, the approximation that $K$ is temperature independent holds, and $k'$ values can therefore be used to compare results obtained using similar methods in different laboratories.

## Column Efficiency

Separation occurs as the compounds in the mixture reach equilibrium between the stationary and mobile phases. With the introduction of the fresh mobile phase, a new equilibrium is reached and there is further transfer of molecules between the stationary and mobile phases. This process occurs over and over again as the mobile phase is forced down the column, facilitating

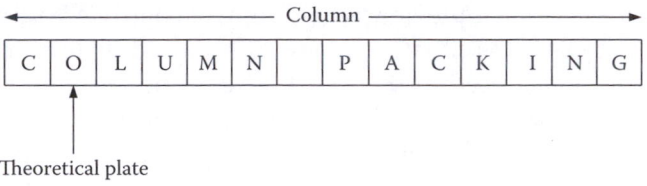

**Figure 2.5**  Theoretical plates.

separation of the mixture of compounds. The distance within the column that represents one of these transfers between stationary and mobile phase is called a *theoretical plate* (see Figure 2.5). The shorter or narrower the theoretical plate is, the more theoretical plates will be present within a given column length. Because the theoretical plate is a measure of the degree of separation, the higher the number present in any given column, the better the degree of separation will be. The longer the column is, the more theoretical plates will be present; therefore, the better the degree of separation will be.

It should be noted, however, that the concept of the theoretical plate is a hypothetical one. It is used to help in understanding the separation processes occurring within the column environment. Theoretical plates can be used as a physical measure of the degree of separation of any given column under a specific set of parameters. This is often referred to as *column efficiency* and can be calculated as follows:

$$N = 5.54 \left( \frac{t_R}{w_{\frac{1}{2}}} \right)^2$$

where $t_R$ is the retention time of the peak of interest, $w1/2$ is the width of the peak at half the height, and $N$ is the number of theoretical plates within a given column. The height of a theoretical plate can also be considered, and this is represented by the following equation:

$$H = \frac{N}{L}$$

**Note:** The greater the number of plates is, the better or more efficient the column is.

The measurement of theoretical plates is not a measure of the separation itself and it is not indicative of whether a column will separate a particular mixture of compounds. These concepts will be dealt with later in the text.

It can be shown that column efficiency decreases with increased retention time. Part of the reason for this is that the longer any sample is in the column, the more diffuse the sample zone becomes, resulting in a widening of the chromatographic peak known as *band broadening*. The reasons for band broadening are many and it is important to understand what these are so that they can be minimised.

### Eddy Diffusion

The mobile phase transports the sample molecules through the chromatographic column. The column itself is packed with a stationary phase made up of small spherical particles typically in the region of 5 μm in size. Some of the molecules in the sample mixture will take the direct route through the packing material (represented by the green squares in Figure 2.6); others, however, will take a lengthier route, weaving in and around the packing material (represented by the red and blue squares in Figure 2.6). As a result of this, the sample plug introduced as a narrow band at the top of the column at the point of injection becomes more and more spread out the longer it remains in the column. This results in a broadening of the compound peak as illustrated in Figure 2.6. Eddy diffusion is independent of mobile phase flow rate through the column.

### Longitudinal Diffusion

As the mobile phase transports the sample molecules through the chromatographic column, the sample molecules will naturally diffuse outward, just as a drop of ink will diffuse out when dropped into a glass of water. This diffusion is known as *longitudinal diffusion* (Figure 2.7) and results in a broadening of the compound peak as illustrated in the figure. The faster the mobile phase is, the less time the sample remains in the column; therefore, there is less effect from longitudinal diffusion.

### Resistance to Mass Transfer

The stationary phase material used in HPLC is made up of very small porous silica particles. The pore structure within each particle will vary in size and

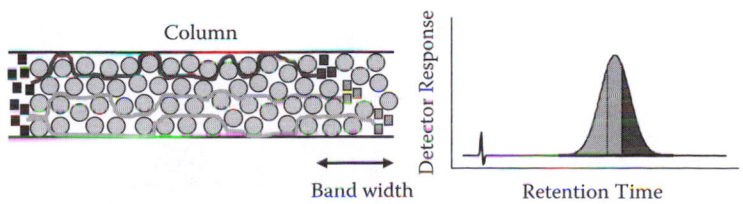

**Figure 2.6 (See colour insert following page 142.)** Eddy diffusion.

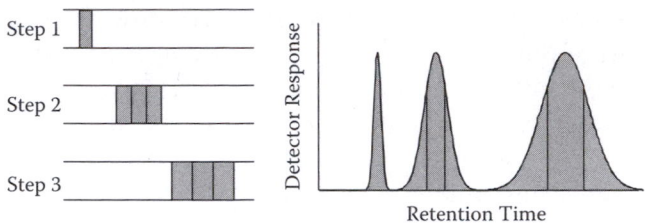

**Figure 2.7** (**See colour insert following page 142.**) Longitudinal diffusion.

shape. As each compound travels down the column, it interacts with the stationary phase.

This process is not instantaneous and a finite time is required for the solute to transfer by diffusion through the mobile phase in order to enter the stationary phase. Thus, the molecules close to the stationary phase will enter it immediately, whereas those some distance from the stationary phase will enter it sometime later. However, as the mobile phase is moving, during the time the molecules are diffusing toward the stationary phase, they will be swept along the column and thus away from those molecules that were closer to the stationary phase and entered it rapidly. This causes the solute peak to be dispersed (spread); the process is called *resistance to mass transfer* in the *mobile phase* (see Figure 2.8).

Dispersion due to resistance to mass transfer in the stationary phase is exactly analogous to that in the mobile phase. Those molecules close to the surface of the stationary phase will leave the surface and enter the mobile phase before those that have diffused further into the stationary phase and require a longer period to diffuse back to the surface. Thus, molecules that quickly enter the mobile phase because they were close to the surface will be swept away from molecules still diffusing to the surface. This process also causes the solute peak to be dispersed; this process is called *resistance to mass transfer* in the *stationary phase*. The flow rate of the mobile phase will have a bearing on the resistance to mass transfer. Slower flow rates will reduce this phenomenon.

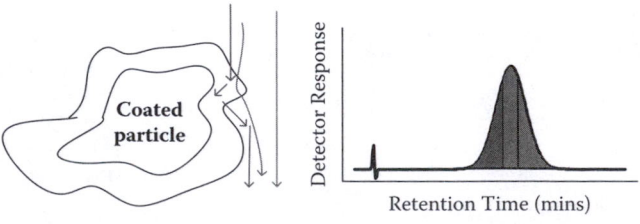

**Figure 2.8** (**See colour insert following page 142.**) Resistance to mass transfer.

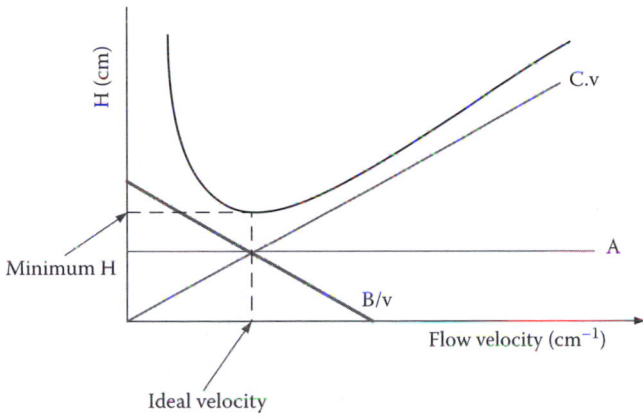

**Figure 2.9** Van Deemter plot.

All of the preceding concepts, although theoretical, can be represented by the Van Deemter equation. The Van Deemter equation and subsequent plot allow for optimisation of the system parameters resulting in efficient separations:

$$H = A + \left(\frac{B}{v}\right) + C.v$$

where $H$ represents the plate height, $v$ represents the flow velocity, and $A$, $B$, and $C$ are parameters representing Eddy diffusion, longitudinal diffusion, and resistance to mass transfer, respectively.

This can be represented in the Van Deemter plot shown in Figure 2.9. The practical applications of this will be discussed later in Chapters 6 and 10.

## Resolution

Resolution ($R$) is a measure of the separation between compounds in a mixture. To achieve optimum chromatographic resolution, the goal is to have what is known as baseline separation, where $R \geq 1.5$. Baseline separation is shown in Figure 2.10.

Resolution of two compounds, $A$ and $B$, can be defined using the following equation, which measures resolution in terms of retention time ($t_R$), peak width ($W$), and peak width at half height ($W_{1/2}$):

$$R = 2\frac{(t_r)_B = (t_r)_A}{W_A + W_B}$$

or

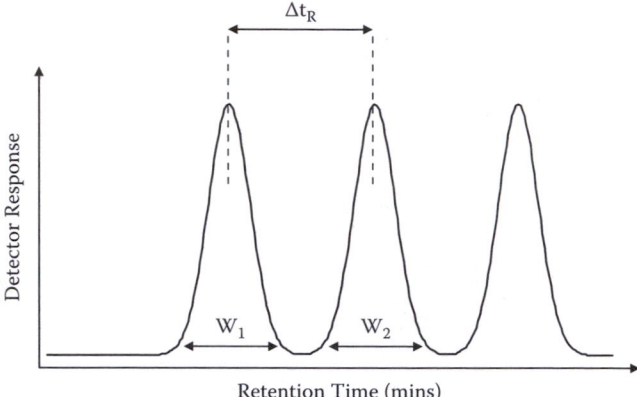

**Figure 2.10** Chromatogram showing baseline separation.

$$R=1.18\frac{\left(t_r\right)_B-\left(t_r\right)_A}{W_{1/2A}+W_{1/2B}}$$

It can also be useful to represent resolution using another equation, which uses the number of theoretical plates ($N$), the separation factor ($\alpha$), and the capacity factor ($k'$) to measure resolution as follows:

$$R=\left(\frac{\sqrt{N}}{4}\right)\cdot\left(\frac{\left(\alpha-1\right)}{\alpha}\right)\cdot\left(\frac{k'_B}{\left(1+k'_B\right)}\right)$$

This equation demonstrates that it is possible to optimise resolution by altering or controlling the terms indicated previously. The higher the number of theoretical plates is, the better the resolution. This can be achieved by lengthening the column or reducing the particle size. Doubling the column length will increase the resolution by a factor of 1.4; however, the analysis time will be doubled, thus increasing the cost per sample. Separation can also be improved by controlling the capacity factor ($k'$). Because the capacity factor is a measure of retention, this can be achieved by altering the composition of the mobile phase or by increasing the column length. The selectivity (separation) factor ($\alpha$) is also a measure of separation, so controlling this will have an effect on the resolution. This can be achieved by

- changing the mobile phase
- increasing or decreasing the column temperature
- changing the column stationary phase
- changing the mode of separation (e.g., moving to ion exchange chromatography)

*Tip: Lengthening the column can lead to increased band broadening.*

## Peak Shape

The simplest description of retention behaviour considered previously suggests that all molecules of a particular analyte should travel down the column at the same rate and emerge at exactly the same time. If this were the case, chromatography would be easy and poor resolution due to overlapping peaks would be very rare.

Thus, we considered a more sophisticated model that explains how band broadening gives rise to peaks with a measurable width. This more realistic model explains why the analyte that needs to be quantified may not always be well resolved from other closely eluting compounds. Even this model, however, needs to be refined further because it suggests that peaks should be symmetrical, while examination of real chromatograms shows that they are often asymmetrical. Some degree of tailing is almost always present, and fronting is not uncommon (see Figure 2.14).

In this section, we will deal with the major contributions to the asymmetry of peaks. We will first deal with asymmetry due to overloading, and then we will deal with asymmetry due to secondary interactions.

### *Asymmetry Due to Overloading*

Previously, it was assumed that the distribution coefficient ($K$) was independent of solute concentration. Thus, as the amount of solute is increased, the concentration in the mobile phase and the concentration in the stationary phase both increase in a linear relationship represented as a straight line graph of $C_{mob}$ against $C_{stat}$, which has slope $K$ (Figure 2.11). This is often referred to as a linear sorption isotherm. Because the line is straight, the value of $K$ is always the same, regardless of the concentration in the mobile phase.

This simple description is, of course, not realistic, and there are two significant deviations from this ideal behaviour. At high levels of solute, one of

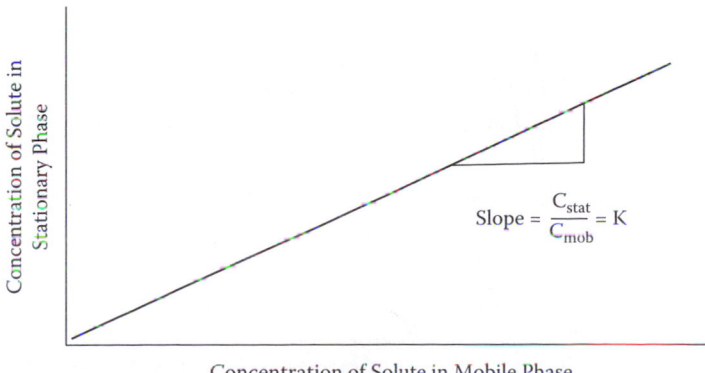

$$\text{Slope} = \frac{C_{stat}}{C_{mob}} = K$$

Concentration of Solute in Stationary Phase

Concentration of Solute in Mobile Phase

**Figure 2.11** Ideal relationship between $C_{stat}$ and $C_{mob}$.

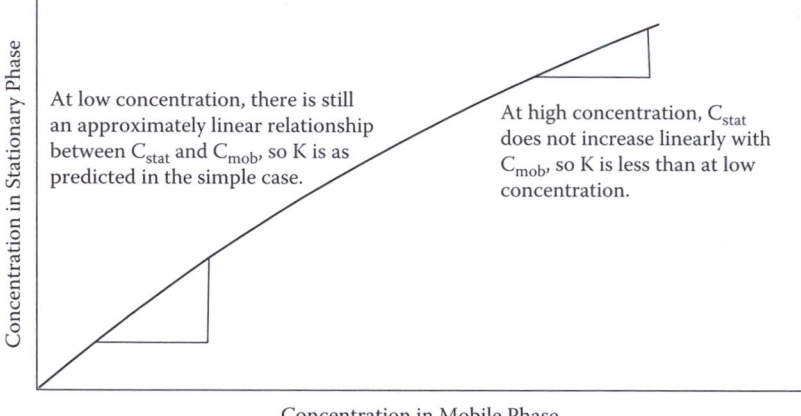

At low concentration, there is still an approximately linear relationship between $C_{stat}$ and $C_{mob}$, so K is as predicted in the simple case.

At high concentration, $C_{stat}$ does not increase linearly with $C_{mob}$, so K is less than at low concentration.

Concentration in Stationary Phase

Concentration in Mobile Phase

**Figure 2.12** Langmuir sorption isotherm.

the phases (usually the stationary phase) can start to become saturated with solute. Thus, if the concentration of solute in the mobile phase is increased, the accompanying increase in concentrations in the stationary phase is not as great as would be predicted. Thus, the simple relationship between $C_{stat}$ and $C_{mob}$ breaks down, and the linear sorption isotherm becomes a Langmuir sorption isotherm (Figure 2.12). If $K$ is measured at different points on the graph, it will be found to be less at high values of $C_{mob}$ than at low values of $C_{mob}$. Because the retention behaviour of solutes depends on $K$, this obviously means the solute will travel more quickly when it is present at high concentration than at low concentration.

Alternatively, there may be cooperative absorption of the analyte so that, as the amount of analyte increases, the concentration in the stationary phase increases more rapidly than would otherwise be predicted. The plot of $C_{mob}$ against $C_{stat}$ consequently curves upward in a shape known as an anti-Langmuir sorption isotherm (Figure 2.13). This is less common than saturation of the stationary phase, but as some solutes (e.g., some acids) are absorbed in the stationary phase, they alter its nature so that they have greater affinity for it (or solubility in it). This means that $K$ is higher at high values of $C_{mob}$ and the solute will therefore travel more slowly than at low concentration.

Let us consider the impact of this on chromatography by imagining ourselves in three different situations travelling down an HPLC column along with a sample of solute that has just been injected onto the column. As the band-broadening effects described previously start to become obvious, a Gaussian distribution of solute concentrations starts to develop. If there is no effect of concentration on $K$ (a linear sorption isotherm), then all that happens is that the band continues to broaden symmetrically as it travels down the column. A symmetrical Gaussian peak will elute from the column.

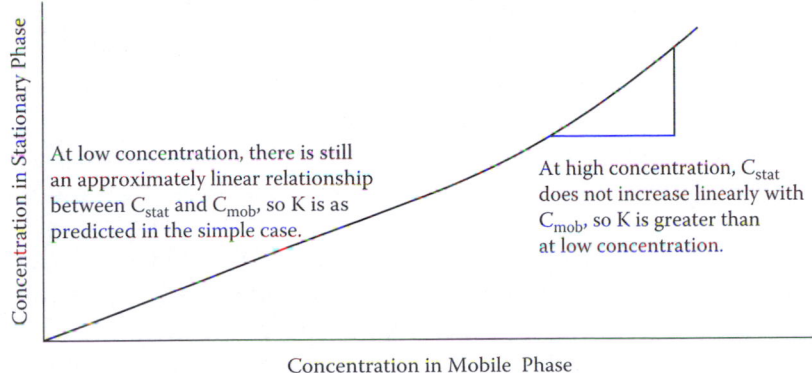

**Figure 2.13** Anti-Langmuir sorption isotherm.

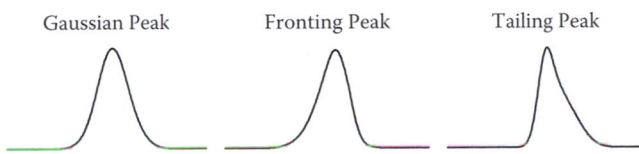

**Figure 2.14** Peak shape.

In the second situation, however, a Langmuir sorption isotherm applies, so the value of $K$ at the centre of the band, where concentrations are high, may be lower than it is at the leading or trailing edges. This means that molecules of solute at the centre of the peak will spend more time in the mobile phase (perhaps because the stationary phase is saturated) and will therefore travel more quickly than those at either edge. This means they will leave the slower molecules at the trailing edge behind and catch up with those at the leading edge. This will give rise to a tailing peak shape (Figure 2.14). This is a very common peak shape when large samples are introduced, and it can cause peaks due to otherwise well separated analytes to overlap, thus making quantification difficult.

In the third situation, an anti-Langmuir isotherm applies, and $K$ is higher at the centre of the band than at either edge, perhaps due to some form of cooperative absorption. This means that molecules in the centre of the band spend more time in the stationary phase and thus travel more slowly than molecules at the leading or trailing edge. Thus, the trailing edge catches up with the centre of the band, while the leading edge gets away. This gives rise to a fronting—even a jagged—peak shape (Figure 2.14). Although less common than tailing, it is also a significant problem at high sample loads.

The answer to both of these problems (fronting and tailing) is discussed in Chapter 10.

### Asymmetry Due to Secondary Interactions

As has been discussed previously, a number of different interactions give rise to retention on HPLC columns that can be exploited to separate analytes. In a well designed column, normally one major type of interaction will be exploited, and other interactions will be suppressed in order to keep separations simple, predictable, and efficient. A good example of this is traditional RP HPLC, in which the main interactions are hydrophobic interactions between solutes and a nonpolar, C18 derivatised silica stationary phase.

In some columns, however, there may be additional (polar/polar or H-bonding) interactions with exposed –OH groups on the silica. This will give rise to peak tailing of polar solutes because a small proportion of solute molecules will be additionally retained by these secondary interactions, causing them to lag behind the main band. In new columns, the number of secondary sites may be low and the amount of tailing may be unnoticeable, but as the column ages, some of the C18 residues may be lost, exposing additional –OH groups and increasing the amount of tailing. Some strategies to deal with this problem are considered in Chapter 10.

An additional source of secondary sites may be derived from irreversibly retained components from previous samples. These may form a contaminated layer at the top of the column rich in secondary sites and may give rise to significant tailing. Again, strategies for dealing with this problem are considered in Chapter 10.

### Measurement of Peak Tailing (Asymmetry)

Peak tailing is a measure of the peak symmetry. In an ideal world, all peaks would be Gaussian in shape or fully symmetrical; unfortunately, the real world is somewhat different and peak shapes usually have an element of peak tailing (see Figure 2.15). Quite often this is due to the nature of the packing material, the chromatographic system, and/or the compounds being separated. Singly or in combination, these factors can lead to peak tailing and this can be expressed using the asymmetry factor $(A_s)$,

$$A_S = \frac{b}{a}$$

which is calculated by dividing the width of the back portion of the peak $(b)$ by the width of the front portion of the peak $(a)$ at 10% of the peak height. $a$ and $b$ are measured from the leading or trailing edge of the peak, respectively, to a line dropped perpendicular to the apex of the peak (see Figure 2.16).

A symmetrical peak would have an asymmetry factor equal to 1.0. Most column manufacturers would consider a peak tailing factor of 0.9–1.2 to be acceptable. An unsymmetrical peak ($A_s > 1.2$ or $A_s < 0.9$) can have a detrimental effect on resolution and on the calculation of peak area for quantitative

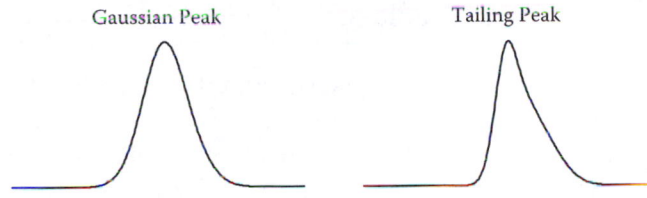

**Figure 2.15**  Peak tailing.

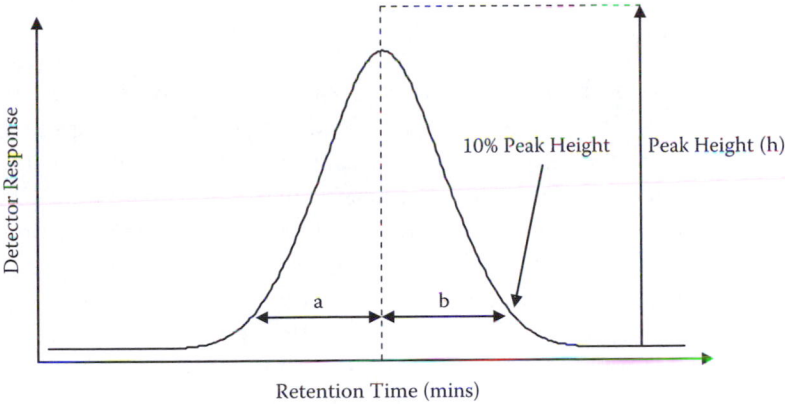

**Figure 2.16**  Peak symmetry.

analysis. Steps can be taken to reduce peak tailing and these will be discussed further in Chapter 10.

Peak asymmetry can also be measured using the *U.S. Pharmacopoeia* (USP) definition, which defines the first section of the peak as *f* and the peak width as the value at 5% of the peak height (see Figure 2.17):

$$T = \frac{w_{0.05}}{2f}$$

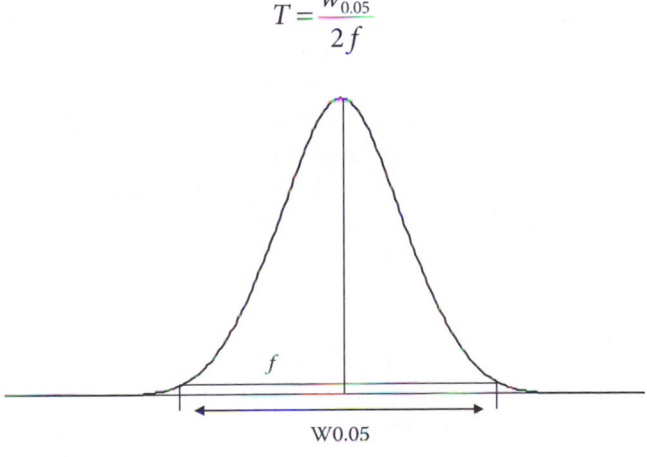

**Figure 2.17**  USP tailing.

## Chemical Bonding and Polarity

Before embarking on the theory of polarity and the solubility of compounds in a solvent, it is necessary to review briefly basic chemistry and the different types of chemical forces that might be at work. A number of different forces may be at work within a single molecule (intramolecular forces) and between different individual molecules in a compound (intermolecular forces).

### Basic Chemistry

To start, let us discuss the nature of chemical bonding. First, consider the periodic table (Figure 2.18). In this table, the elements are divided into groups (denoted by the numbers at the tops of the columns); these group numbers tell us the number of valence electrons that are associated with the element (Table 2.2).

Each atom of an element has protons, neutrons, and electrons associated with it. Consider the electrons only, because they are involved in bonding, and look at how they are accommodated. Atoms are composed of orbitals, which can be thought of as 'boxes' in which the electrons are stored. There are four different types of orbitals: s-, p-, d-, and f-orbitals.

An s-orbital is spherical in shape and can hold only two electrons. When the first s-orbital is filled, we can move on to the second s-orbital, which is bigger than the first. The third is bigger again and so on. As the number of electrons increases, we move up from a 1s shell to a 2s shell (again, spherical and accommodating two electrons) and then to a 2p and then a 3s. p-Orbitals are different in shape from an s-orbital; instead of only one, we have three: $p_x$, $p_y$, and $p_z$. The p-orbitals are shown in Figure 2.19.

s-Orbitals can accommodate two electrons; p-orbitals can accommodate six electrons (two in the $p_x$-orbital; two in the $p_y$-orbital, and two in the $p_z$-orbital). d-Orbitals can accommodate 10 electrons (2 in each of the following orbitals: $d_{xy}$, $d_{xz}$, $d_{yz}$, $d_{x^2-y^2}$, $d_{z^2}$). f-Orbitals can accommodate 14 electrons in seven orbitals but are more complicated and rarely involved in the bonding between organic elements covered in this textbook. These electrons fill up the shells according to the Aufbau principle: The electrons will fill up the shells in a particular order, based on the energy of the orbital; the lowest energy orbitals fill up first (see Figure 2.20).

### Octet Rule

This rule states that elements will react in order to achieve eight valence electrons. This corresponds with the electronic configuration of noble gases and is the result of the s- and p-orbitals being filled (with the exception of He). Elements try to achieve noble gas configuration due to the stability that having eight valence electrons affords.

Key:

| Atomic number | Symbol | Atomic weight |
|---|---|---|
| 1 | H | 1.0079 |

| 1 | 2 | 3 | 4 | 5 | 6 | 7 | 8 | 9 | 10 | 11 | 12 | 13 | 14 | 15 | 16 | 17 | 18 |
|---|---|---|---|---|---|---|---|---|---|---|---|---|---|---|---|---|---|
| 1 H 1.0079 | | | | | | | | | | | | | | | | | 2 He 4.003 |
| 3 Li 6.941 | 4 Be 9.0122 | | | | | | | | | | | 5 B 10.811 | 6 C 12.011 | 7 N 14.007 | 8 O 15.999 | 9 F 18.998 | 10 Ne 20.1798 |
| 11 Na 22.989 | 12 Mg 24.3051 | | | | | | | | | | | 13 Al 25.982 | 14 Si 28.086 | 15 P 30.974 | 16 S 32.065 | 17 Cl 35.453 | 18 Ar 39.948 |
| 19 K 39.0983 | 20 Ca 40.078 | 21 Sc 44.956 | 22 Ti 47.867 | 23 V 50.942 | 24 Cr 51.996 | 25 Mn 54.938 | 26 Fe 55.845 | 27 Co 58.938 | 28 Ni 58.6934 | 29 Cu 63.546 | 30 Zn 65.546 | 31 Ga 69.723 | 32 Ge 72.64 | 33 As 74.922 | 34 Se 78.96 | 35 Br 79.904 | 36 Kr 83.798 |
| 37 Rb 85.4678 | 38 Sr 87.62 | 39 Y 88.906 | 40 Zr 91.224 | 41 Nb 92.906 | 42 Mo 95.94 | 43 Tc [98] | 44 Ru 101.07 | 45 Rh 102.906 | 46 Pd 106.42 | 47 Ag 107.868 | 48 Cd 112.411 | 49 In 114.818 | 50 Sn 118.710 | 51 Sb 121.760 | 52 Te 127.60 | 53 I 126.904 | 54 Xe 131.293 |
| 55 Cs 132.9054 | 56 Ba 137.328 | 57–71 lanthanides | 72 Hf 178.49 | 73 Ta 180.948 | 74 W 183.84 | 75 Re 186.207 | 76 Os 190.23 | 77 Ir 192.217 | 78 Pt 195.084 | 79 Au 196.967 | 80 Hg 200.59 | 81 Tl 204.383 | 82 Pb 207.2 | 83 Bi 208.980 | 84 Po [209] | 85 At [210] | 86 Rn [222] |
| 87 Fr [223] | 88 Ra [226] | 89–103 actinides | 104 Rf [261] | 105 Db [262] | 106 Sg [266] | 107 Bh [264] | 108 Hs [277] | 109 Mt [268] | 110 Ds [271] | 111 Rg [272] | | | | | | | |

| 57 La 138.905 | 58 Ce 140.116 | 59 Pr 140.908 | 60 Nd 144.242 | 61 Pm [145] | 62 Sm 150.36 | 63 Eu 151.964 | 64 Gd 157.25 | 65 Tb 158.925 | 66 Dy 162.500 | 67 Ho 164.930 | 68 Er 167.259 | 69 Tm 168.934 | 70 Yb 173.04 | 71 Lu 174.967 |
|---|---|---|---|---|---|---|---|---|---|---|---|---|---|---|
| 89 Ac [227] | 90 Th 232.038 | 91 Pa 231.036 | 92 U 238.029 | 93 Np [237] | 94 Pu [244] | 95 Am [243] | 96 Cm [247] | 97 Bk [247] | 98 Cf [251] | 99 Es [252] | 100 Fm [257] | 101 Fm [257] | 102 Md [258] | 103 Lr [262] |

**Figure 2.18** Periodic table.

**Table 2.2  IUPAC Numbers for the Periodic Table**

| IUPAC Number | Name | Valency |
|---|---|---|
| 1 | Alkali metals | 1 |
| 2 | Alkaline earth metals | 2 |
| 13 | | 3 |
| 14 | | 4 |
| 15 | | 3 |
| 16 | | 2 |
| 17 | Halogens | 1 |
| 18 | Noble gases | 0 |

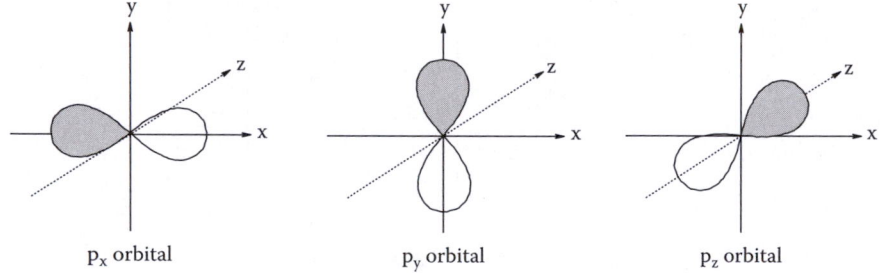

**Figure 2.19**  p-Orbitals.

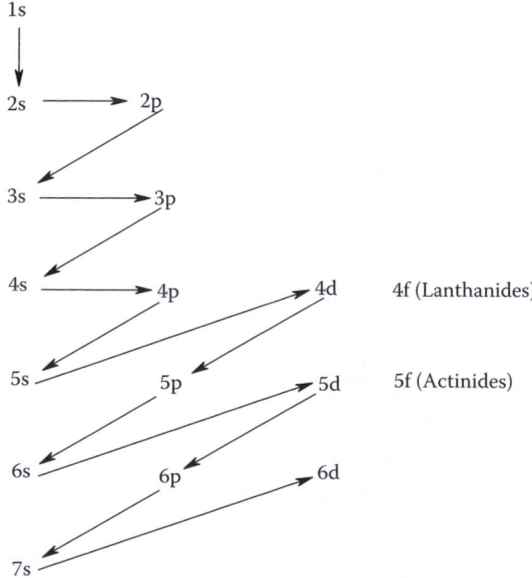

**Figure 2.20**  Order of orbital shell filling.

## Electron Configuration and Valency

The periodic table (Figure 2.18) shows that elements with the same number of valence electrons are grouped together: Group 1 elements have one electron in the valence shell, group 2 elements have two electrons in the valence shell, group 3 elements have three electrons in the valence shell, and so on. The group number in the periodic table tells us how many electrons are in the valence shell and how many electrons we have for bonding. The availability for bonding (i.e., the number of spaces available) is called *valency*.

Valency is the ability of an element to bond with other elements; the valency number tells us how many bonds can be formed. For example, consider hydrogen (Figure 2.21), which is found in group 1 of the periodic table. This element has one electron present in the first s-orbital (denoted by $1s^1$). An s-orbital is spherical in shape and can have a maximum of two electrons present in the shell; therefore, hydrogen can bond with only one other element. The further along the periodic table we move, the more electrons we have present in the element, which means that we need more shells to accommodate them.

Consider the first three rows of the periodic table (Figure 2.22). As can be seen, the table starts to fill up with every element, due to the addition of an electron. When we reach the end of the row (when we reach group 18 elements), it is usual to denote this as a helium shell [He], neon shell [Ne], argon shell [Ar], and so on. For example, if we consider the electron configuration of chlorine (Cl), we can denote this as having a neon shell plus the additional electrons in the 3s and 3p shells. This is written as $[Ne]3s^23p^5$.

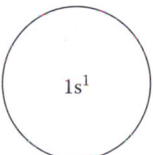

**Figure 2.21** $1s^1$ orbital.

| H<br>$1s^1$ | | | | | | | He<br>$1s^2$ |
|---|---|---|---|---|---|---|---|
| Li<br>$1s^22s^1$ | Be<br>$1s^22s^2$ | B<br>$1s^22s^22p^1$ | C<br>$1s^22s^22p^2$ | N<br>$1s^22s^22p^3$ | O<br>$1s^22s^22p^4$ | F<br>$1s^22s^22p^5$ | Ne<br>$1s^22s^22p^6$ |
| Na<br>$1s^22s^2$<br>$2p^63s^1$ | Mg<br>$1s^22s^2$<br>$2p^63s^2$ | Al<br>$1s^22s^2$<br>$2p^63s^23p^1$ | Si<br>$1s^22s^2$<br>$2p^63s^23p^2$ | P<br>$1s^22s^2$<br>$2p^63s^23p^3$ | S<br>$1s^22s^2$<br>$2p^63s^23p^4$ | Cl<br>$1s^22s^2$<br>$2p^63s^23p^5$ | Ar<br>$1s^22s^2$<br>$2p^63s^23p^6$ |

**Figure 2.22** First three rows of the periodic table.

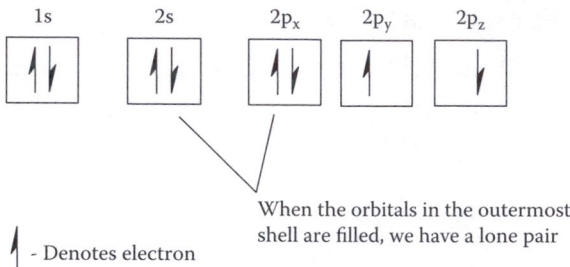

When the orbitals in the outermost shell are filled, we have a lone pair

$\uparrow$ - Denotes electron

**Figure 2.23** Electrons-in-boxes notation for oxygen.

The outermost shell of the atom (in the preceding example of chlorine, the outermost shell is the 3p-orbital, which contains five electrons) is called the *valence shell* and the electrons contained within this shell are called the *valence electrons*. These electrons have the highest amount of energy in the atom and are therefore the most reactive. For this reason, these electrons will participate in bonding and chemical reactions. All other electrons within the atom generally do not participate in such reactions.

It is important to know where electrons are to be found in a molecule to understand the properties and reactivity that it will have. Oxygen has an electron configuration of $1s^2 2s^2 2p^4$; this means that there are two electrons in the spherical 1s-orbital; two electrons in the spherical, although longer than the 1s, 2s-orbital; and four electrons distributed among the three 2p-orbitals (see Figure 2.23). Orbitals can be represented as boxes, where the electrons are denoted by arrows with quantum theory assigning a 'spin': This is why we have two arrows: one in the up direction and the other in the down position.

The reason that we find two electrons in separate p-orbitals ($2p_y$ and $2p_z$) is that electrons do not want to share the space, unless they really have to. Electrons, in this manner, are like humans on a bus: No one wants to sit next to someone unless there are no other spaces available. The 'space' that is left in the orbitals means that oxygen has a valency of two, which means that oxygen can bond with two other elements (with a valency of one) or form a double bond with one other element.

We continue with oxygen as an example and look at the position of the four electrons in the p-orbitals (Figure 2.24). We know that the electrons are present in the orbitals, but they are in continual motion in what is known as an electron cloud. Because the electrons move around in the electron cloud, it is unlikely for all electrons to be found on the same side of an atom or molecule. Although electrons are in continual random motion, we represent them as being present in particular orbitals for ease of explanation and understanding.

Let us now consider the types of bonding that can occur between atoms of elements. These are formed through *intermolecular forces* or *intramolecular forces*.

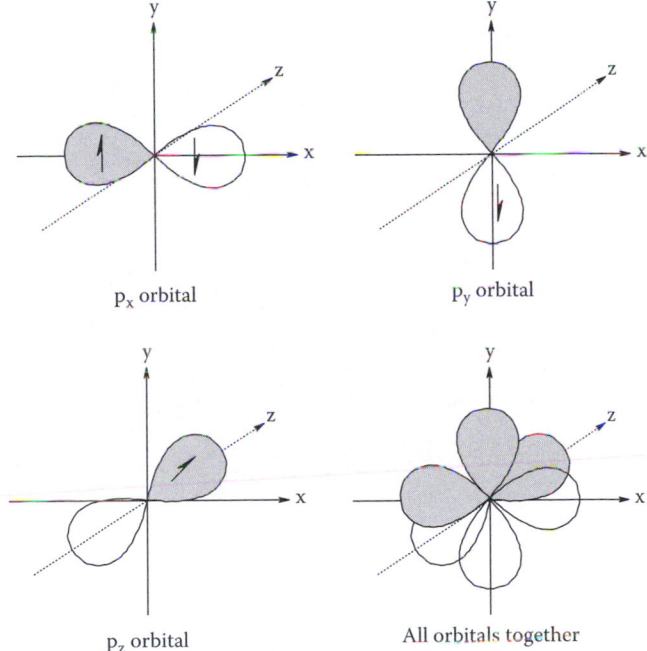

**Figure 2.24** p-Orbitals.

## Intramolecular Forces

Intramolecular forces occur within a molecule and include ionic and covalent bonds.

### Ionic Bonding

Ionic bonds are formed by an interaction between charged elements (e.g., sodium fluoride—NaF). This type of bonding occurs between metals and nonmetals. The bond is a result of the positive charge on the *cation* (in the preceding example, this is the sodium—$Na^+$) and negative charge of the *anion* (in the example, this is the fluoride ion—$F^-$).

Sodium has one valence electron; this means that it has a valency of one fluorine, has seven valence electrons, and also has a valency of one. Because sodium has one less electron than required to fill the 3s-orbital, it 'wants' to obtain noble gas electron configuration (all filled shells); sodium will therefore release its valence electron, giving it an overall positive charge. Fluorine, on the other hand, has seven valence electrons; it too 'wants' to achieve noble gas electron configuration but cannot release seven electrons. It is easier and requires less energy to accept an electron from another atom (which is willing to release an electron). Figure 2.25 shows the valence electrons of sodium and fluorine.

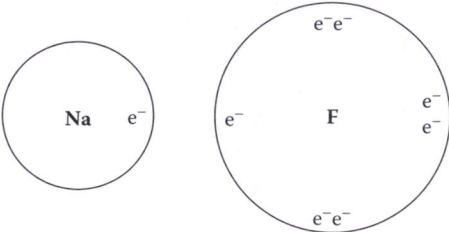

**Figure 2.25** Sodium fluoride.

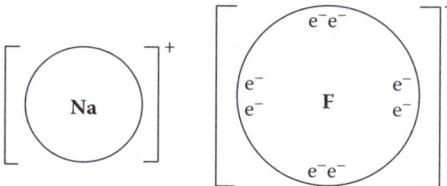

**Figure 2.26** Electron configuration.

The valence electron from the sodium is released to the fluorine (see Figure 2.26). It is the attraction between the positive charge on the cation (sodium) and the negative charge on the anion (fluorine) that forms the ionic bond.

## Covalent Bonding

This type of bonding occurs between nonmetals. Instead of a bond formed from the attraction between a cation and an anion (as with ionic bonds), we have the formation of a bond from the sharing of electrons between atoms.

Let us use oxygen as an example of covalent bonding: Oxygen atoms have six valence electrons (oxygen is found in group 6 of the periodic table) and therefore have a valency of two. Oxygen atoms can bond with each other by sharing their electrons to achieve noble gas electron configuration (see Figure 2.27). It is not possible for electrons to be released in this situation (unlike with ionic bonding) because one of the atoms must lose electrons, resulting in an overall positive charge, and the other accepts the electrons to have a negative charge. Therefore, the oxygen atoms share the electrons, resulting in a double bond (O=O) (see Figure 2.28).

## Intermolecular Forces

These forces occur between molecules rather than between atoms in a molecule. There are both weak and strong intermolecular forces; the weaker of these forces are the Van der Waals dispersion forces and dipole–dipole moments. The strongest of these forces is hydrogen bonding.

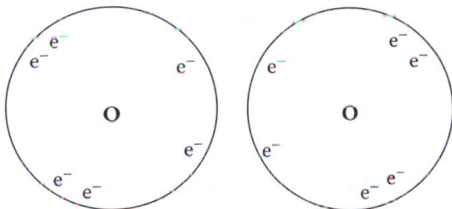

**Figure 2.27** Oxygen atoms.

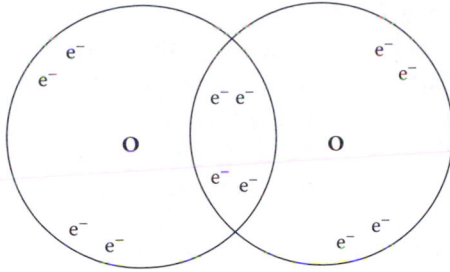

**Figure 2.28** Bonding of oxygen atoms.

### Van der Waals Dispersion Forces

These forces are sometimes called London dispersion forces after Fritz London, who suggested how they work, and form the weakest type of bonding that occurs between molecules. However, they are present in all substances.

In a previous section, we looked at the electron configuration and valency concept that electrons present within an electron cloud are in continual random motion. Because the electrons may collect at one end of a molecule, we can find an instantaneous or fleeting dipole moment; this results in a molecule having a temporary positive or negative charge at certain times. It is these charges that will be attracted to each other and are termed Van der Waals forces. These are the only types of forces present in nonpolar, covalent molecules.

### Dipole–Dipole Moments

These forces occur between the positive end of a polar molecule and the negative end of another. These positive and negative charges are not full charges, but rather are partial charges and are denoted as such ($\delta^+$ and $\delta^-$, respectively). Because these charges are localised and permanent, they effectively cancel each other out and thus have no overall effect on the net charge on the molecule. However, due to the existence of these partial charges, the sides of molecules with opposite charges will be attracted to one another.

For example, with HCl molecules, a dipole moment is present because the hydrogen atom has a slightly positive charge ($\delta^+$) and the chlorine has a slightly negative charge ($\delta^-$). A dipole–dipole interaction takes place between

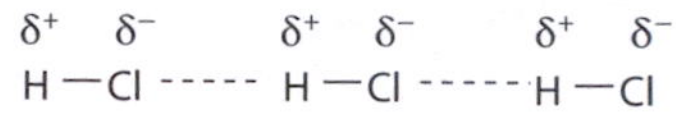

$$\begin{array}{ccccc} \delta^+ \quad \delta^- & & \delta^+ \quad \delta^- & & \delta^+ \quad \delta^- \\ H-Cl & ----- & H-Cl & ----- & H-Cl \end{array}$$

**Figure 2.29** Dipole-dipole interaction.

oppositely charged sides of the HCl molecules that are adjacent to each other (see Figure 2.29):

## Hydrogen Bonding

These forces are strong dipole–dipole interactions that occur between molecules with electron-rich groups such as –OH, –NH, or –FH. In these groups, the small hydrogen atoms are bonded to highly electronegative atoms (O, N, and F), which induce electrons toward them. This results in strong partial charges on both the electronegative atom and the hydrogen ($\delta^+$ and $\delta^-$, respectively). Water is a very good example of where hydrogen bonding exists: The oxygen atoms have two lone pairs resulting in an overall $\delta^-$ charge with the hydrogen having the $\delta^+$ charge (see Figure 2.30).

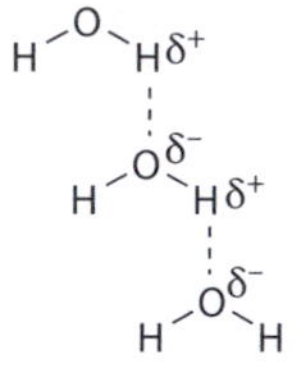

**Figure 2.30** Hydrogen bonding between water molecules.

## Mixed Intermolecular Forces

The intermolecular forces have been described as separate entities; however, in the types of samples with which forensic scientists have to deal, it is rare to find only one force at work. When the intra- or intermolecular forces are discussed, we consider the interaction between molecules of identical natures (i.e., the hydrogen bonding between water molecules or dipole–dipole interactions between $CO_2$ molecules). However, because of the complex matrices that are forensic science samples (containing many substances), we must consider all types of forces that may be present.

## Polarity of Compound and Solubility

### Electronegativity and Polarity

Electronegativity is the ability of an atom to draw electrons in a covalent bond toward itself. In terms of electronegativity and the periodic table, we find that, moving across the periodic table (from left to right), electronegativity increases and, moving down a group, electronegativity decreases. This means that, of all of the elements in the periodic table, fluorine is the most electronegative. A small section of the elements of the periodic table with associated electronegativity values is shown in Figure 2.31.

| H<br>2.1 | | | | | | | | He<br>-- |
|---|---|---|---|---|---|---|---|---|
| Li<br>1.0 | Be<br>1.5 | | B<br>2.0 | C<br>2.5 | N<br>3.0 | O<br>3.5 | F<br>4.0 | Ne<br>-- |
| Na<br>0.9 | Mg<br>1.2 | | Al<br>1.5 | Si<br>1.8 | P<br>2.1 | S<br>2.5 | Cl<br>3.0 | Ar<br>-- |

**Figure 2.31** Electronegativity and the periodic table.

If we return to the bonding of atoms and consider this in relation to the electronegativity of the atoms involved, we find that the larger the difference in the electronegativities, the stronger the bond. Polarity, on the other hand, refers to the intermolecular forces between the $\delta^+$ end of a molecule and the $\delta^-$ end of another molecule of the same or a different molecule.

### Polar and Nonpolar Bonds

Polar covalent bands are formed by sharing electrons, as with covalent bonding; however, the electrons are not shared equally. The charged 'ends' of the bonds are called dipoles. The end of the bond with the largest electronegativity will be the slightly negatively charged end, whereas the end with the smallest electronegative charge will be the slightly positively charged end. Nonpolar covalent bonds are formed by the equal sharing of electrons. Examples include hydrogen, chlorine, oxygen, and other diatomic elements.

What does this mean for a forensic scientist? Molecules may contain polar covalent bonds and may or may not be polar; the overall polarity of the molecule is determined by measuring the dipole moment. This dipole moment will depend upon the distance between the two ends (oppositely charged) of the bond and also on the overall degree of separation of charge between the atoms in the bond. If we find that a molecule contains equal polar bonds that balance each other around the central atom, we find that the molecule will be nonpolar, even though the bonds within it are polar.

Consider, for example, water ($H_2O$) (Figure 2.32); this molecule has an unequal sharing of electrons between the central oxygen atom and the two hydrogen atoms and therefore is polar. On the other hand, methane ($CH_4$) has an equal sharing of electrons and therefore is nonpolar. However, the solvents and substance being analysed by liquid chromatography are generally referred to in terms of their polarity in comparison to each other.

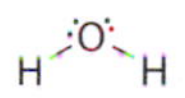

**Figure 2.32** Water molecule.

Because we now know that the polarity of a compound will affect the separation that might occur in a mixture, it follows that the solubility of

**Table 2.3   Polarity (P) of Common Solvents**

| Solvent | Polarity Index (P) |
| --- | --- |
| Pentane | 0 |
| Hexane | 0.1 |
| Heptane | 0.1 |
| Cyclohexane | 0.2 |
| Toluene | 2.4 |
| 2-Propanol | 3.9 |
| n-Butanol | 4 |
| Tetrahydrofuran (THF) | 4 |
| Chloroform | 4.1 |
| Acetone | 5.1 |
| Methanol | 5.1 |
| Acetonitrile | 6.2 |
| Water | 10.2 |

those compounds in a solvent will also be affected by polarity. Not only do we have to consider the polarity of the compounds in which we are interested, but we also need to consider the best solvent for dissolution of these compounds.

When solvents are used in liquid chromatography, it is useful to know their polarity in relation to each other so that an appropriate choice of mobile phase or mobile phase combination might be chosen. Table 2.3 shows a list of commonly used solvents in mobile phases in order of increasing polarity.

Molecules are polar if they have a net dipole moment. This occurs if they contain an unsymmetrical arrangement of polar bonds; therefore, it is necessary to consider multiple aspects of the molecule when deciding whether it will be polar or nonpolar. It is necessary to think about the polarity of the individual bonds in the molecule as well as the size and shape of the molecule. To make this a little easier, we can consider these factors by looking at the following five stages:

1. Begin by looking for permanently charged groups; these are very polar and inevitably will make the molecule polar. One should also consider the pKa of groups that may become charged at certain pH values (e.g., –COOH, –NH$_2$, or phenolic –OH groups).
2. The next step is to look for individual bonds in the molecule that might be expected to be polar. If the difference is more than 0.5, it can be assumed that the bond is to some extent polar. Very large differences (>1.7) suggest that the bond will be ionic and likely to dissociate in aqueous solutions.
3. Having now established which bonds are polar and which are not, it is necessary now to determine whether the bonds are symmetrical

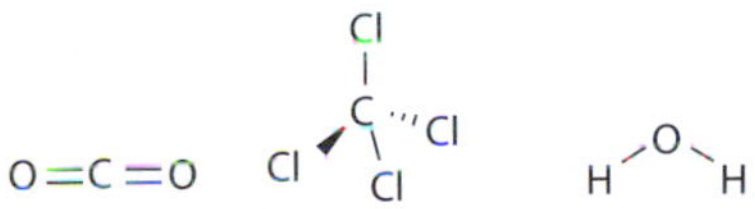

**Figure 2.33** Molecular geometry.

or not. To do this, the shapes of the molecules in three dimensions, rather than just in two, must be known because they are commonly drawn in texts. To illustrate this with an example, consider carbon dioxide ($CO_2$), which is nonpolar (dipole moment = 0) because the polar C=O bonds (difference in electronegativity = 1.0) are arranged opposing each other in a straight line (see Figure 2.33). $H_2O$ is polar (dipole moment = 3.0 D) because the O–H bonds are arranged at an angle of 104.5° (see Figure 2.33). Taking a slightly more complicated example, carbon tetrachloride ($CCl_4$) is nonpolar (dipole moment = 0) because the four polar C–Cl bonds (difference in electronegativity = 0.5 D) are arranged in a tetrahedron shape with bond angles of 109.5° (see Figure 2.33).

4. The next step to remember is that a small polar region may be outweighed by a large nonpolar region, leading to the molecule being overall nonpolar (or at least not very polar). Methanol, for example, contains a polar O–H bond (difference of electronegativity = 0.4 D), a polar C–O bond (difference of electronegativity = 1.0), and three nonpolar C–H bonds (difference of electronegativity = 0.4). The molecule is not symmetrical and methanol is polar (dipole moment = 2.9 D)—less polar than water, but polar nonetheless. However, as we proceed through the series of straight chain alcohols, we find that they become progressively less polar as the molecule becomes dominated by the increasing number of nonpolar C–H and C–C bonds (difference of electronegativity = 0). Thus, octanol ($C_8H_{17}OH$), with a dipole moment of 1.65 D, is nonpolar. This effect can be even more marked when dealing with tertiary molecules.

5. Finally, it is necessary to be aware of the notion of 'polarisability', which can be very important in determining chromatographic behaviour. To explain this, it is useful to consider that some molecules contain loosely held electrons that can be dragged away from their normal position when a molecule is close to a strong dipole or charged species. An important example is the electron clouds that can be found above and below an aromatic ring. These can be attracted to positive charges on another molecule, or the $\delta^+$ end of a molecule with a significant dipole. This induces a dipole in the aromatic molecule that can lead to a reasonably strong dipole–dipole interaction.

**KEY POINT SUMMARY**

Retention factor:

$$k' = \frac{t_R - t_M}{t_M}$$

Separation factor:

$$a = \frac{k'_2}{k'_1}$$

Column efficiency:

$$N = 5.54 \left( \frac{t_R}{W_{\frac{1}{2}}} \right)^2$$

Theoretical plate height:

$$H = \frac{N}{L}$$

Van Deemter equation:

$$H = A + \left( \frac{B}{v} \right) + C.v$$

Resolution equations:

$$R = 2\frac{\left(t_r\right)_B - \left(t_r\right)_A}{W_A + W_B}$$

or

$$R = 1.18\frac{\left(t_r\right)_B - \left(t_r\right)_A}{W_{1/2A} + W_{1/2B}}$$

$$R = \left( \frac{\sqrt{N}}{4} \right) \cdot \left( \frac{(\alpha - 1)}{\alpha} \right) \cdot \left( \frac{k'_B}{\left(1 + k'_B\right)} \right)$$

## Peak shapes:

Peak shapes:

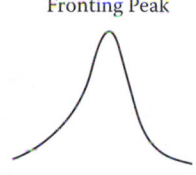
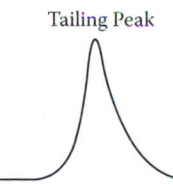

## Asymmetry factor:

$$A_S = \frac{b}{a}$$

## Filling order of electron shells:

Filling order of electron shells:

1s

2s      2p

3s      3p

4s      4p      4d      4f (Lanthanides)

5s      5p      5d      5f (Actinides)

6s      6p      6d

7s

- Intermolecular forces:
  - ionic
  - covalent
- Intramolecular forces:
  - Van der Waals
  - dipole–dipole
  - hydrogen bonding

## QUESTIONS

*Worked example:* Using the following chromatogram, calculate the following parameters:

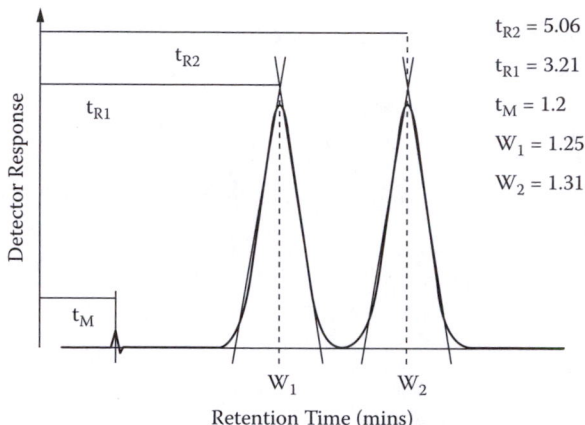

$t_{R2} = 5.06$

$t_{R1} = 3.21$

$t_M = 1.2$

$W_1 = 1.25$

$W_2 = 1.31$

Calculate $k'$ for peaks 1 and 2.

*Solution*

$$k' = \frac{t_R - t_M}{t_M}$$

Peak A:

$$k' = \frac{3.21 - 1.20}{1.20} = 1.68$$

Peak B:

$$k' = \frac{5.06 - 1.20}{1.20} = 3.22$$

Calculate the resolution between peaks 1 and 2.

*Solution*

$$R = \frac{2\left[\left(t_r\right)_B - \left(t_r\right)_A\right]}{W_A + W_B}$$

$$R = \frac{2\left(5.06 - 3.21\right)}{1.25 + 1.31} = 1.45$$

Calculate α for peaks 1 and 2:

$$\alpha = \frac{k'_2}{k'_1}$$

$$\alpha = \frac{3.22}{1.68} = 1.92$$

Calculate the number of theoretical plates for peak 2 ($W1/2 = 0.94$):

$$N = 5.54 \left( \frac{t_R}{W_{\frac{1}{2}}} \right)^2$$

$$N = 5.54 \left( \frac{5.06}{0.94} \right)^2 = 161$$

1. The following chromatogram represents the separation of four common drugs of abuse. The column packing material is C18 and the mobile phase is methanol/water 50/50. Calculate the following parameters:
   a.  $k'$ for peaks 1–4
   b.  α for the best resolved and the worst resolved pair
   c.  Resolution for the following peaks pairs: 1/2, 2/3, 3/4
   d.  Plate number for peaks 1–4

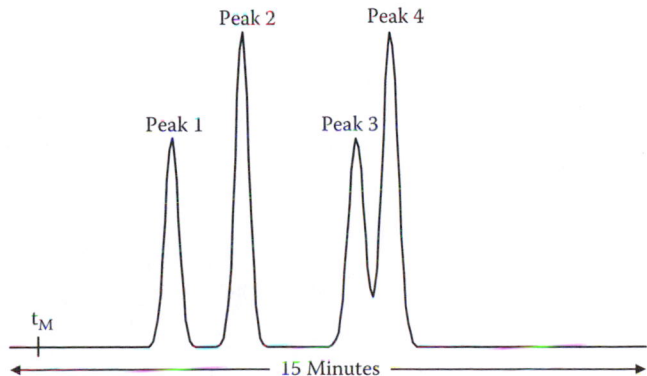

2. A resolution of 2.5 is required in the separation of two compounds X and Y. How many theoretical plates must the column have in order to ensure that the resolution criterion is met? $k'_A$ and $k'_B$ are 1.54 and 2.38, respectively?

$$R = \left(\frac{\sqrt{N}}{4}\right) \cdot \left(\frac{(\alpha-1)}{\alpha}\right) \cdot \left(\frac{k'_B}{\left(1+k'_B\right)}\right)$$

3. Calculate the tailing factor $T$ using the USP method and the 10% peak height method for the following peak.

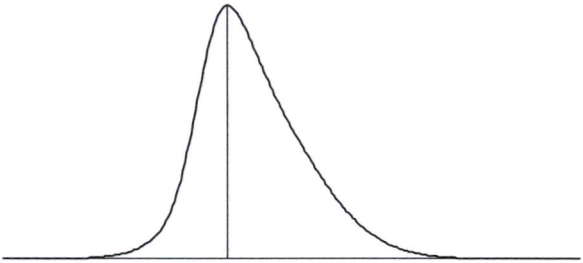

4. Using the Van Deemter equation, what would be the impact on $H$ of a large resistance to mass transfer ($c$) value?
5. Why is the flow velocity inversely proportional to the longitudinal diffusion but directly proportional to resistance to mass transfer?
6. Give the electron configuration for the following atoms:
   a. Nitrogen
   b. Boron
   c. Phosphorus
   d. Chlorine
   e. Bromine
7. For the previously mentioned atoms, use the electron-in-a-box notation.
8. Using the following chemical structure of $\Delta^9$-tetrahydrocannibinol (the active constituent of cannabis), identify the potential sites for inter- and intramolecular interactions:

CH$_3$

OH

H$_3$C—
H$_3$C        O        C$_5$H$_{11}$

$\Delta^9$-THC

9. Of the following three molecules, which is the most polar and why?

Morphine                    Atropine                    Cannabidiol

10. Explain why water is a more polar solvent than pentane.

## Further Reading

### Books/Online Publications

Cazes, J., and R. Scott. 2002. *Chromatography theory.* Boca Raton, FL: CRC Press.

Cole, M. D. 2003. *The analysis of controlled substances,* New York: John Wiley & Sons.

Dong, M. W. 2002. *Modern HPLC for practicing scientists.* New York: Marcel Dekker Inc.

Heftmann, E. 2004. *Chromatography: Fundamentals and techniques.* New York: Elsevier.

Higson, S. 2004. *Analytical chemistry.* New York: Oxford University Press.

Jönsson, J. A. 1987. *Chromatographic theory and basic principles.* Boca Raton, FL: CRC Press.

Kenkel, J. V. 2003. *Analytical chemistry for technicians,* 3rd ed. Boca Raton, FL: CRC Press.

McMaster, M. C. 2007. *HPLC: A practical user's guide,* 2nd ed. New York: Wiley-Interscience.

Meyer, V. R. 2005. *Practical high-performance liquid chromatography.* New York: John Wiley & Sons.

Shibamoto, T. 1998. *Chromatographic analysis of environmental and food toxicants.* Boca Raton, FL: CRC Press.

Skoog, D. A., S. R. Crouch, F. J. Holler, and D. M. West. 2004. *Fundamentals of analytical chemistry,* 8th ed. Pacific Grove, CA: Thomson-Brooks/Cole.

Waters Corporation. 2009. *How does high-performance liquid chromatography work?* (www.waters.com).

Waters Corporation. 2009. *What is HPLC (high-performance liquid chromatography)?* (www.waters.com).

### Journals

*Journal of Forensic Sciences,* American Academy of Forensic Sciences, American Society for Testing and Materials, Synergy (online service).

# Mobile Phase Preparation, the Use of Buffers, and Sample Preparation

3

## Mobile Phase Preparation

The basic principles surrounding the preparation of mobile phase for HPLC analysis are, on the whole, independent of the separation mechanism used. The HPLC column and sample type will determine the mobile phase choices discussed in Chapter 6. All of the solvents and chemicals used in the preparation of mobile phases should be of a sufficiently high quality depending on the nature of the separation. General purpose reagents are available as analytical reagent (AR) or laboratory reagent (LR) grade; however, for HPLC there is a need for a higher quality reagent for use as a solvent and as a chemical reagent used in the preparation of buffers. Most solvents and chemicals are available as 'HPLC grade', which generally means that they have been through at least one additional purification step; this might include submicron filtration, glass distillation, and rigorous specification testing. An example of the type of information and testing that might be expected to be included in the accompanying certificate of analysis for HPLC grade materials is given in Table 3.1.

It should be noted that different manufacturers use different processes to purify their materials and while all might claim to be of the same standard, quite often they are not. Poor-quality solvents and chemicals may contain impurities and particulate matter that can cause blockages within the system either in the tubing or in the column or detector cell. Other factors to consider are the UV cut-off levels of the solvents being used and the stability. The solvent must have a lower UV cut-off wavelength than the absorption wavelength of any of the compounds of interest within the sample matrix. This is particularly important at shorter wavelengths because the solvent may absorb UV light along with the sample, resulting in a cumulative effect that will adversely affect any quantitative measurement made at that wavelength.

Most solvents are stable but some, like tetrahydrofuran (THF), have a very short shelf life. This should be noted at the time of preparation and an expiry date assigned to the mobile phase that does not exceed the expiry date for the THF. UV cut-off levels for some commonly used solvents are given in Table 3.2.

45

**Table 3.1    Example of a Certificate of Analysis for HPLC Grade Solvent**

## Company Name and Logo

### Certificate of Analysis

| | | |
|---|---|---|
| Product: Methanol | Grade: HPLC | Lot/Batch Number: MN6759H67a |
| Description: Clear colourless mobile liquid | | Date of Analysis: 3 June 2009 |

Application: HPLC Expiry Date:
Use within 2 years of opening

| Test | Result | Units | Current Specification |
|---|---|---|---|
| Absorbance @ 200 nm | 0.08 | A.U. | ≤0.2 |
| Absorbance @ 210 nm | 0.026 | A.U. | ≤0.05 |
| Absorbance @ 220 nm | 0.01 | A.U. | ≤0.02 |
| Absorbance @ 230 nm | 0.005 | A.U. | ≤0.01 |
| Absorbance @ 240 nm | 0.002 | A.U. | ≤0.005 |
| Absorbance @ 250 nm | 0 | A.U. | ≤0.005 |
| Acidity/alkalinity (meq/g) | 0.00005 | meq/g | ≤0.00015 |
| Residue after evaporation (ppm) | 0 | ppm | ≤2 |
| Water | 0.0028 | % | ≤0.01 |

Additional information:

Signature of QA laboratory manager:

Date:

**Table 3.2    UV Cut-off Data**

| Solvent | UV Cut-off (nm) |
|---|---|
| Acetone | 330 |
| Acetonitrile | 210 |
| Methanol | 210 |
| Water | <190 |
| Isopropyl alcohol | 210 |
| Acetic acid | 235 |
| Triethylamine | 275 |

All of the components for mobile phase preparation can be purchased from reputable or approved suppliers and any changes in supplier should be checked thoroughly before materials are used. Additional validation of the method may be required when there has been a change in supplier of a critical reagent. All mobile phases should be filtered and degassed prior to use. Further discussion in relation to mobile phase quality is given in Chapter 12.

## Filtration

Filtration is required to remove any particulate matter that could cause damage to the pump seals and could block the column and narrow tubing if they are not removed. The choice of filter will depend on the level of particulate contamination and the mobile phase composition. An example of this might be the use of a 0.45 µm nylon membrane filter, which is compatible with most mobile phases. Information in relation to filter compatibility for commonly used mobile phase components for RP-HPLC is given in Table 3.3.

Let us consider what might happen if we used a PVC filter with an acetonitrile-based mobile phase. We have already discussed the 'like dissolves like' principle in Chapter 2 as the basis for a chromatographic separation and the same processes apply in the case of the filter and the mobile phase. If the polarities of both are similar, then we would expect one to dissolve in the other. In the case of the PVC filter and acetonitrile, we would find that the filter begins to disintegrate as the mobile phase is filtered through. This is not an acceptable situation because it introduces impurities into the mobile phase and destroys the filter. It should be noted that a large number of different types of filters are available for different applications and one should always adhere to the manufacturer's advice regarding their use.

**Table 3.3   Filter Compatibility Table**

| Filter Material | Acetic Acid | Acetonitrile | Formic Acid (25%) | Hydrochloric Acid (1 $M$) | Methanol | THF | Ammonia (1 $M$) |
|---|---|---|---|---|---|---|---|
| Cellulose acetate | √ | x | √ | √ | x | x | √ |
| Nylon | √ | √ | √[a] | √ | x[b] | √ | √ |
| Polypropylene (PP) | √ | x | √ | √ | √ | x | x[b] |
| PVC (polyvinyl chloride) | √ | x | x[b] | √ | x[b] | x | x[b] |
| PTFE | x | √ | x | √[a] | √ | √ | √ |

*Note:* This table has been compiled using several published compatibility tables.

√  = recommended for use.

x  = not recommended for use.

√[a] = not recommended for use at higher concentrations.

x[b] = check filter efficiency before use; may cause cracking and swelling.

## Degassing

All mobile phases should be degassed prior to use. This removes any dissolved gases that have accumulated on standing or as a result of the mixing or dispensing processes. Air bubbles can cause 'spiking' on the chromatogram baseline (see Chapter 10) and can be disruptive to the chromatography process and subsequent integration. Some HPLC systems have a built-in facility to degas the mobile in the solvent delivery system. Some manufacturers offer a system that allows the solvent to pass through a membrane tube, which is held at constant vacuum and is permeable to gases but not to liquids. The gas is then discharged from the surface of the membrane under the vacuum condition.

Despite the automated degassing, it is still advisable to filter each of the components of the mobile phase under vacuum before use. Degassing can be performed manually using a combination of vacuum filtration (see Chapter 3 for filter compatibility information), followed by immersion in an ultrasonic bath for a short period of time. Care should be taken when using an ultrasonic bath because of the generation of heat that can alter the mobile phase composition by evaporation of the more volatile organic solvent.

Another effective method that can be used to degas the mobile phase is helium sparging. A stream of helium is piped through the mobile phase via a sparge frit situated at the bottom of the mobile phase container. The helium, which is not soluble in most HPLC solvents, drives dissolved air out of the mobile phase. There is a small risk that the composition of the mobile phase may change slightly with continual use of the helium sparge. This can be reduced by ensuring that the helium flow rate is kept to a minimum.

## Mixing Mobile Phase Components

### Single-Component Mobile Phase

The desired quantity of solvent should be measured out using a measuring cylinder or beaker, filtered and degassed, and then placed into a suitable HPLC solvent reservoir. The vessel should be large enough to hold the volume of solution but should not be too big so that evaporation might occur.

### Dual-Component Mobile Phase

In this case, the mobile phase can be prepared manually (premixed) or it can be mixed automatically using the HPLC solvent delivery system.

**Manual Mixing**   When the mobile phase is to be prepared manually, the organic portion and the aqueous portion should be independently measured out using measuring cylinders or by weight where appropriate. Both phases should then be added together in a separate HPLC solvent bottle and mixed by inversion (when possible). This is the most accurate way to prepare a dual-component mobile phase manually. There are alternative ways to prepare mobile phases manually, but these tend to lead to variation and inaccurate results (see Chapter 10). The mobile phase should then be filtered and degassed under vacuum prior to use.

**Automated Mixing**    Binary HPLC systems have a built-in mixing facility that allows two components to be mixed together accurately without the requirement for premixing manually. In a high-pressure system, two dual-pistons operate in tandem to draw in the individual components of the mobile phase. The mobile phase then passes through a pulse damper, into a mixing chamber, through a purge valve, and onto the column.

**Nomenclature**    It is common practice to record mobile phase mixtures by stating the organic portion first in the following format:

- 50/50 methanol/water would represent a mobile phase containing 50% methanol and 50% water.
- 40/60 acetonitrile/water would represent a mobile phase containing 40% acetonitrile and 60% water.
- 10/90 methanol/water containing 0.1% trifluoroacetic acid would represent a mobile phase containing 10% methanol and 90% water containing 0.1% trifluoroacetic acid.

### HYPOTHETICAL EXAMPLE 3.1

To prepare 1,000 mL of a mobile phase containing 70% methanol and 30% water, measure out 700 mL methanol and 300 mL water in separate measuring cylinders. Mix both components together; mix thoroughly and filter and degas before use.

## Multicomponent Mobile Phase Mixing

It is more common with multicomponent mobile phases to utilise the mixing facility within the quaternary pump. The quaternary pump operates by means of a proportioning valve and a single pump that draws the individual components from the solvent bottles into an inlet valve. The piston pushes the mobile phase into a damper, through a purge valve, and onto the column. This is an example of a low-pressure quaternary system. A high-pressure system would operate using four individual pump heads to draw the components into the system, rather like the binary system described previously. Most manufacturers supply low-pressure quaternary systems, most likely due to cost implications, ease of maintenance, and size.

Multicomponent mixtures are often complex in nature and subtle changes in composition can have a dramatic effect on the chromatography (see Chapter 6). Therefore, it is considered more reproducible and accurate to use the instrument to mix the components rather than relying on the operator, where the potential for error is considered to be greater. Each individual component should be filtered and degassed and then placed in an appropriate solvent reservoir. It is common practice to record multicomponent mobile phase mixtures in the following format:

50/10/40 methanol/acetonitrile/water or
50% methanol/10% acetonitrile/40% water

## Buffers

### Acids/Bases—pKa

According to the Brønsted–Lowry theory for acids and bases, the following definitions stand:

An acid is a substance that is a proton ($H^+$) donor.
A base is a substance that is a proton acceptor.

This can be represented by the equation in Figure 3.1.

At equilibrium, the equilibrium for strong acids (HA) is displaced to the right-hand side (Equation i); alternatively, for strong bases ($A^-$), the equilibrium is displaced to the left-hand side. It follows from this that the stronger an acid is, the weaker its conjugate base will be, and the stronger the base is, the weaker its conjugate acid will be. For reference, Table 3.4 shows the relative strength of common acids and bases.

The ability of an acid (HA) to donate a proton (see Figure 3.2) depends on the polarity of the H–A bond ($H^{\delta+}$–$A^{\delta}$) (more polar bonds dissociate more easily into ions) and also on the strength of the H–A bond (weaker bonds dissociate more readily).

Dissociation of an acid, as shown in Equation i, can be described by an equilibrium constant ($K_a$):

$$K_a = \frac{[H^+][A^-]}{[HA]}$$

A ⇌ $H^+$ + B

Acid     Proton     Base

**Figure 3.1** Equivalent constraint.

**Table 3.4   Relative Strengths of Common Acids and Bases**

| Decreasing Acid Strength | Acid | Base | Increasing Base Strength |
|---|---|---|---|
| | $H_2SO_4$ | $HSO_4^-$ | |
| | HI | $I^-$ | |
| | HCl | $Cl^-$ | |
| | $HNO_3$ | $NO_3^-$ | |
| | HF | $F^-$ | |
| | $H_2O$ | $OH^-$ | |
| | $NH_3$ | $NH_2^-$ | |

$$HA \xrightleftharpoons{K_a} H^+ + A^-$$

**Figure 3.2**  Acid dissociation.

To find the acid dissociation constant, we can take the negative log of both sides of the equation:

$$-\log K_a = -\log\left(\frac{[H^+][A^-]}{[HA]}\right)$$

This can then be simplified to give

$$-\log K_a = -\log[H^+] + \left(-\log\frac{[A^-]}{[HA]}\right)$$

We can use the following information to help us to simplify the equation further: pKa = $-\log K_a$ and pH = $-\log$ [H$^+$]. This simplification gives us

$$pKa = pH - \log\frac{[A^-]}{[HA]}$$

This in turn can be rearranged to give us the following equation, which is known as the Henderson–Hasselbalch equation:

$$pH = pK_a + \log\frac{[A^-]}{[HA]}$$

where *HA* is undissociated acid and *A*$^-$ is conjugate base.

## Buffer Selection and Preparation

From our discussion in Chapter 2, we know that separation of a mixture of compounds is based on their affinity with the stationary and mobile phases within the system. In reversed phase chromatography, the affinity is based on the polarity of the different components and will vary depending on whether the compounds favour a polar mobile phase (i.e., are more polar in nature) or a nonpolar stationary phase (i.e., are more nonpolar in nature). The separation is measured in terms of retention time and the retention time is heavily dependent on the mobile phase pH.

When analysing ionic or ionisable compounds, it is recommended that the pH of the mobile phase be controlled. A small change in mobile phase pH can have a significant effect on the chromatography and hence the separation. Therefore, it is necessary to control the pH of the mobile phase in order to maintain and sometimes promote an adequate separation. This can be achieved by the addition of an aqueous buffer to the mobile phase mixture.

### *Basic Compounds*

A base is a compound that can accept hydrogen ions ($H^+$). A compound with a pKa greater than about 13 is considered to be a strong base. At pH levels below the pKa, a basic compound will assume a positive charge ($R-NH_3^+$). This will result in the compound becoming ionised and behaving more like a polar molecule, resulting in reduced retention. At pH levels above the pKa, the compound will assume a neutral charge ($R-NH_2$) and will become more hydrophobic in nature—hence, an increase in retention time. Operating at pH levels around the pKa can result in varied retention times due to subtle changes in the pH values and unpredictability in relation to the charge on the compound.

In Figure 3.3, we can see that at pH levels approaching the pKa, both the ionised and non-ionised species will be present. It is better to operate in a pH region where only one of the forms exists. Operating under high pH conditions where the compounds are in the non-ionised state can lead to failure of the stationary phase (see Chapter 6) and can also mean that nonpolar basic compounds, in particular, can become irreversibly stuck to the stationary phase. In cases where this might be a problem, it is often better to buffer the mobile phase at a much lower pH, where all of the compound is present in the ionised state, leading to shorter but more stable retention times.

### *Acidic Compounds*

An acid is a compound that donates a hydrogen ion ($H^+$) to another compound. A strong acid completely dissociates in water. A weak acid, on the other hand, only partially dissociates in water and, at equilibrium, both

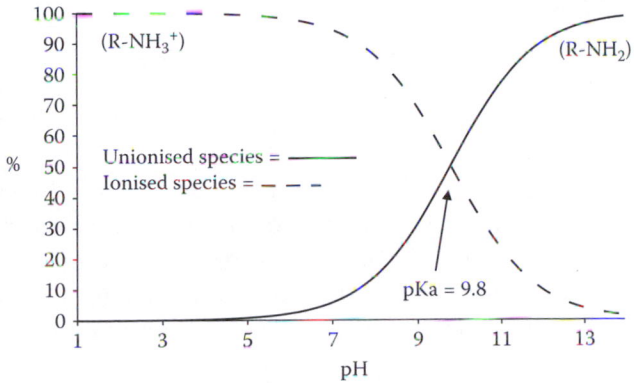

**Figure 3.3** Ionisation of a weak base with changing pH.

the acid and the conjugate base are in solution. At pH levels above the pKa, acidic compounds will lose a hydrogen ion and will assume a negative charge (R-COO⁻) and will behave more like a polar molecule resulting in reduced retention. At pH levels below the pKa, the compound will assume a neutral charge (R-COOH) and will become more hydrophobic in nature—hence, an increase in retention time.

In Figure 3.4, we can see that at pH levels approaching the pKa, both the ionised and non-ionised species will be present. As with the basic compounds, it is better to operate in a pH region where only one of the forms exists. For stronger acids, this can mean operating under very high pH conditions, which can lead to failure of the stationary phase (see Chapter 10). In cases where this might be a problem, it is often better to choose a different chromatographic separation mode, such as ion chromatography.

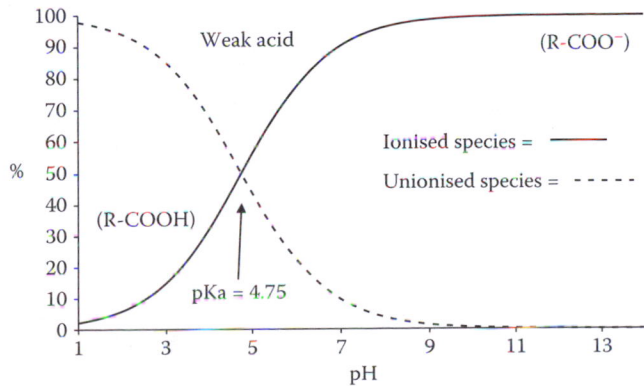

**Figure 3.4** Ionisation of a weak acid with changing pH.

## Buffer Selection

Optimum buffering capacity (the ability to resist small changes in pH upon the introduction of a sample at a different pH) occurs at a pH equal to the pKa of the buffer. In general, it can be expected that most buffers will provide adequate buffering capacity for controlling mobile phase pH within ±1 pH unit of the buffer pKa. Beyond that, buffering capacity will be inadequate. Table 3.5 lists some commonly used buffers for reversed phase HPLC. Included in the table are pKa values and optimal ranges for controlling pH for the buffers listed. Because it is becoming more common to find HPLC interfaced to mass spectrometers, volatile buffers for LC/MS applications are also indicated.

Mistakes can occur when trying to control the pH of the mobile phase and it is not unusual to find methods where the buffer choice and pH range are completely incompatible with the intended separation and there is little or no buffering capacity. For example, it is quite common to see that the aqueous portion of the mobile phase has been adjusted to pH 7 with acetic acid with the expectation that the solution is therefore buffered at pH 7. This is not the case and can lead to confusion. What this means is that the pH of the solution is at 7 before the introduction of the sample. Upon addition of the sample, the pH of the solution may change; this introduces many unnecessary complications that may lead to an increased level of unpredictability in terms of retention time and separation from run to run.

When the buffer is chosen, it is necessary to ensure that the pH range is within ±1 pH unit of the pKa value quoted for that buffer. Buffer selection can involve a substantial amount of method development work and it is

**Table 3.5   Commonly Used Buffers for Reversed Phase HPLC**

| Buffer | pKa | Buffer Range |
|---|---|---|
| Trifluoroacetic acid (TFA)[a] | 0.5 | <1.5 |
| Phosphate | 2.1 | 1.1–3.1 |
|  | 7.2 | 6.2–8.2 |
|  | 12.3 | 11.3–13.3 |
| Formate[a] | 3.8 | 2.8–4.8 |
| Acetate[a] | 4.8 | 3.8–5.8 |
| Citrate | 3.1 | 2.1–4.1 |
|  | 4.7 | 3.7–5.7 |
|  | 5.4 | 4.4–6.4 |
| Tris | 8.3 | 7.3–9.3 |
| Borate | 9.2 | 8.2–10.2 |
| Ammonia[a] | 9.2 | 8.2–10.2 |
| Diethylamine | 10.5 | 9.5–11.5 |

[a]  A volatile buffer suitable for LC-MS.

important to try a number of different buffers and pH values until the optimum one is found for the desired separation.

The pH of the buffer should be selected depending on the pKa values of the components in the mixture and the separation requirements. For any given separation system, there is an attraction between the solute (sample compound) and the stationary and mobile phases. This attraction can be influenced through ionisation or ion suppression of the acidic and basic compounds undergoing separation. Using reversed phase HPLC as an example where there is a nonpolar stationary phase and a polar mobile phase, it would be expected that a nonpolar basic compound would have a high affinity for the stationary phase.

As already demonstrated, the more nonpolar the compound is, the greater will be the affinity for the stationary phase. However, this may be an unattractive proposition in terms of analysis time, mobile phase cost, etc., so, for strongly bound compounds, it may be advantageous to make them less nonpolar in nature. This can be achieved by promoting ionisation of the compound of interest by altering the pH of the solution to below the pKa for the basic compound (see Figure 3.5).

Doing this effectively reduces the attraction for the stationary phase and, in doing so, shortens the retention time. The compound will become charged and acts like a more polar molecule. Amphetamine is a weak base with a pKa of 10.1 and can be used to illustrate this point. As the pH in the mobile phase is reduced, the amphetamine begins to ionise and we would expect to see a reduction in retention time. As the drop in pH approaches the pKa, the change in retention time is quite marked until such time as we reach a state where all of the amphetamine is fully ionised. At this point (pH 2–3), small changes in pH will have less of an effect on the retention time. In summary, for basic compounds, the retention time will decrease as the pH of the mobile phase is decreased.

The reverse also applies when the basic compounds may not be resolved sufficiently. It may be desirable to increase the retention time of the compounds to improve the separation. In this case, it is necessary to maximise the affinity for the stationary phase; this can be achieved by adjusting the

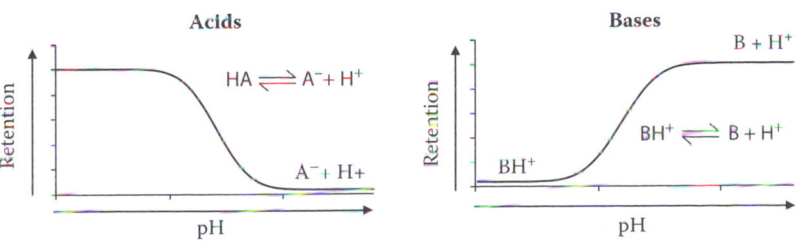

**Figure 3.5** Effect of pH on retention time.

pH of the mobile phase above the pKa of the compound. The pH required to maintain the non-ionised form of very strongly basic compound can present problems due to the lack of stability of the stationary phase at extremes of pH (see Chapter 6).

Separation of organic acids, on the other hand, would normally be performed using ion suppression at pH values of about 2–3; however, at extremes of pH, the non-ionised form of the compound becomes less soluble in water ('like dissolves like'); that is, in the case of an organic acid at low pH (2–3), we have the non-ionised form (less polar) of the acid and a polar solvent, water. When this happens, we would expect to see a reduction in solubility, which can lead to precipitation of the organic acid—especially in the presence of higher concentration of organic modifier in the mobile phase. This can lead to blockage of the HPLC system due to particulate undissolved material in the sample matrix. For acids, the retention time decreases as the pH of the mobile phase is increased (see Figure 3.5).

In cases where extremes of pH are required to effect an adequate separation, a modified column (capable of withstanding extremes of pH) should be used or ion chromatography might present a suitable alternative separation mechanism.

### HYPOTHETICAL EXAMPLE 3.2

Let us look at a separation involving two acidic compounds with pKa values of 4.2 and 4.7. Consider what would happen if we were to

- add no buffer to the mobile phase
- buffer the mobile phase at pH 7
- buffer the mobile phase at pH 3

If we start with a mixture of water and methanol in our mobile phase and do not buffer the system, then we would expect to have a pH somewhere in the region of about pH 7, depending on the exact proportions of methanol and water. When we introduce our acid compounds, we would expect the pH of the system to change and we would effectively have a small pH gradient happening in our sample plug. This would result in poor chromatography represented by inconsistent retention times and broad tailing peaks.

If the mobile phase is buffered at pH 7, then the acids will be ionised; however, the pH has now been controlled within the system. This would result in adequate peak shape and relatively short retention times (depending on exact mobile phase composition), but both of the components are ionised and hence the retention times might still be unpredictable (see Figure 3.6).

If the mobile phase is buffered at pH 3, then both of the acids will be neutral species and hence more hydrophobic. This would result in adequate peak shape, reproducible retention times, and longer retention times. Because both compounds are neutral and more hydrophobic, the affinity for the stationary phase will be much greater. An increase in the organic modifier may help to resolve this issue if the retention time is excessive, but may result in a loss of resolution.

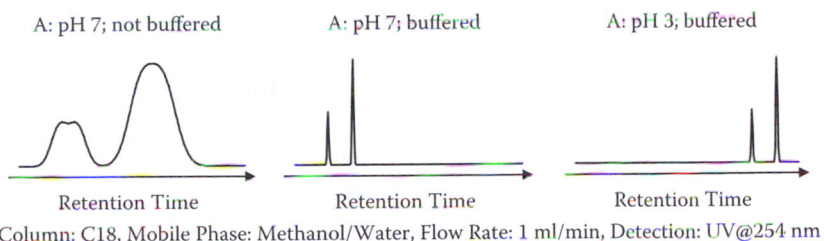

Column: C18, Mobile Phase: Methanol/Water, Flow Rate: 1 ml/min, Detection: UV@254 nm

**Figure 3.6** Effect of pH on the separation of two acidic compounds.

## *Buffer Concentration*

Buffer concentration in RP-HPLC usually has very little effect on the retention time in a given separation provided that there is sufficient buffer present to control the pH. Too little buffer will result in limited buffering capacity and poor chromatography. Solubility problems can arise when the ionic strength of the buffer is high (100 m$M$), especially in the presence of high proportions of organic modifier. The miscibility of the buffer solution with the full range of expected concentrations of organic modifier should be demonstrated before introducing the mobile phase into the HPLC system.

Buffer solutions are prone to bacterial growth and should be prepared regularly and an expiry date assigned. In order to flush out buffer salts from the system, a mobile phase should be prepared that contains water instead of the buffer element. The system, including the column, should be washed thoroughly with this mixture. The column should then be flushed with organic solvent for storage purposes.

## *Buffer Preparation*

Small changes in mobile phase pH can have a pronounced effect on the separation of acidic and basic compounds. It is necessary, therefore, to ensure that a consistent and reproducible approach is taken when preparing mobile phase buffers. A common approach is to measure an amount of water, equivalent to approximately half the final volume, into a suitably sized beaker. The buffer salt should be accurately weighed in a weighing boat and quantitatively transferred to the beaker with the aid of small quantities of water. The pH of the solution should be measured and adjusted, if necessary, at this time. The solution should then be quantitatively transferred into a suitable volumetric flask and made to volume with water. The solution should be inverted several times to ensure thorough mixing. The organic portion of the mobile phase should be dispensed and added to the buffer solution in a suitable HPLC vessel to produce the final mobile phase. This phase should be thoroughly mixed, filtered, and degassed prior to use.

## Sample and Standard Preparation

The extent to which a sample has to be prepared for analysis will be dependent on the nature of the sample that is presented for analysis. For example, a single bag of white powder will require minimal preparation and can be sampled, diluted in an appropriate solvent, mixed, filtered, and dispensed into an HPLC vial ready for analysis. However, not all forensic samples are quite so simple and, even when they might appear simple, some sample pretreatment will be required in order to ensure homogeneity. In most forensic laboratories, a diverse range of samples will require HPLC analysis; some examples are given in Table 3.6.

Each of the samples listed in this table will require some sample preparation, which will range from direct immersion in a solvent to a more sophisticated solid phase extraction (SPE) or liquid–liquid extraction (LLE). In almost all cases, it is necessary to isolate the compound or compounds of interest from the sample matrix. In forensic science, this can often prove challenging because sample matrices can be complex in nature and can include drug cutting agents and biological samples such as liver. A simple sample preparation guide is included in Figure 3.7 in the form of a flow chart.

Reference standards, on the other hand, require a more simple preparation regime. The standard of interest is usually prepared by direct immersion in a suitable solvent. It should be noted, however, that it is recommended that at least one reference standard solution be subjected to the same extraction processes that are applied to the sample. This standard solution will act as a quality control (QC) sample in order to check the extraction processes.

Table 3.6    Examples of Types
of Samples That Might Be Found
in a Forensic Environment

| Sample Type | Sample Form |
| --- | --- |
| Drugs | Powders |
| | Tablets |
| | Capsules |
| | Resin |
| | Oil |
| Toxicology | Blood |
| | Urine |
| | Stomach contents |
| | Liver |
| Dyes | Fibres/paper |
| | Liquid/solid |
| Paint | Chips/flakes |
| | Liquid/solid |

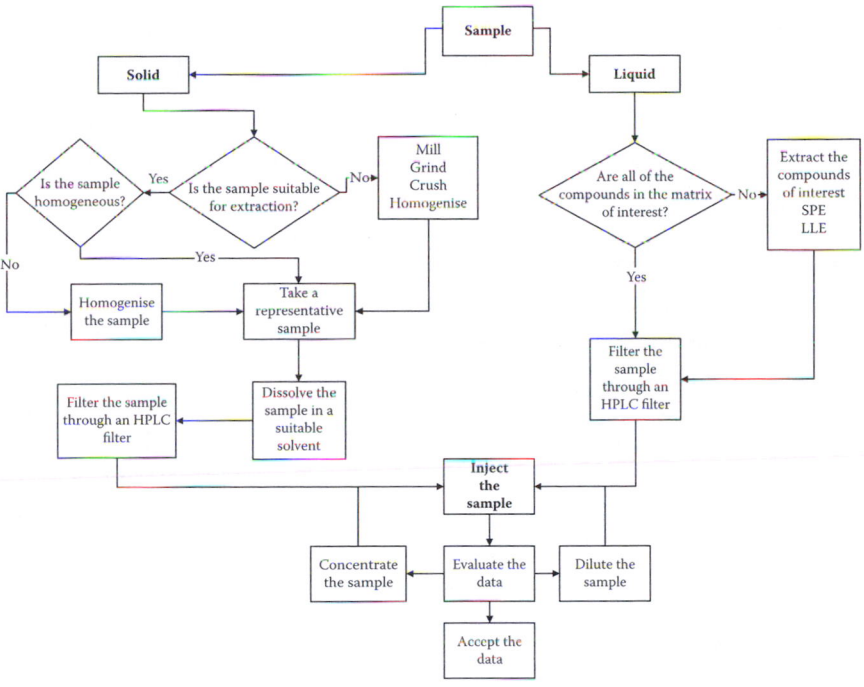

**Figure 3.7** Sample preparation guide.

The extraction process will be discussed in greater detail later in this section under each of the extraction techniques.

Selection of the sample can present a challenge in forensic science, and the method used will depend on a number of factors. If you are presented with a bag containing 1,000 tablets, all of which appear to be approximately the same size and shape, do you test all of the tablets? In terms of the time taken and the cost, the answer is most certainly going to be no. In this case, you will need to be sure that the sample of tablets that you do test is representative of the whole population. The European Network of Forensic Science Institutes (ENFSI) has published a booklet designed to help forensic laboratories decide on a sampling plan for their operations. This booklet, published in 2007, is available free of charge through its Internet site. The United Nations Office on Drugs and Crime has published a similar guide to sampling (2009, available free of charge through its Internet site). Ultimately, the level of sampling applied in any forensic examination will depend on the individual laboratory and the jurisdiction that it serves.

## Sample Preparation Techniques

Many, if not all, samples analysed in forensic science by chromatography require some form of 'clean-up' step for them to be suitable for analysis. With liquid chromatography, samples are required to be 'clean' before they can be introduced into the instrument; if not, it is possible, if not extremely likely, that the column will become clogged with material from the sample matrix. This in turn reduces the sensitivity of the detection, if not impeding analysis altogether. A number of so-called clean-up techniques can be used before chromatographic analysis, depending primarily on the type of sample being analysed.

### Protein Precipitation

Protein precipitation is a technique used in toxicology to remove the protein content of human body fluids and tissues before they are analysed. The reason for this is that these samples of protein content can vary from 6% to more than 50%, by weight, in some tissues. This can greatly affect the possibility of detecting and quantifying drug and metabolite concentrations.

Generally, a precipitation reagent, such as an organic solvent (e.g., acetonitrile or methanol) or a salt and an acid (e.g., ammonium sulphate and hydrochloric acid), will be used. Once the proteins have been precipitated, the solid protein will then be removed by filtering or by centrifugation. The rest of the liquid sample will then be further cleaned up before extraction, or extraction will occur immediately after the protein precipitation stage.

### Extraction Techniques

Extraction techniques—in particular, liquid–liquid extraction and solid phase extraction—are used in toxicological analysis and some drug analysis prior to chromatographic analysis. The process of extraction is used to extract organic substances, such as drugs, directly from body fluids and tissues. The two main types of extraction used in these types of analyses are liquid–liquid extraction and solid phase extraction.

*Liquid–liquid extraction* is a technique in which inert, water-immiscible organic solvents of varying polarity are used to extract the drug substances. Many different solvents can be used, but this will ultimately depend upon the nature of the drugs being extracted—that is, whether the substances are acidic (pKa 1–5), basic (pKa 7–12), or neutral (no acidic or basic functional groups, extractable at all pH levels) molecules. Generally, the extraction will be achieved by agitating the sample with the appropriate solvent in a separating funnel. After varying the pH and agitating or centrifuging the mixture, the solute molecules (the drugs) will be found in the (organic) layer and the inorganic substances will be found in the aqueous layer. In some instances, two immiscible organic solvents will be used in place of the organic/aqueous combination described previously (see Figure 3.8).

This extraction of organic solutes, such as drug substances from biological samples, occurs based on the *Nernst distribution law,* sometimes known as the *partition law.* This law is used to explain the distribution of a solute between two phases.

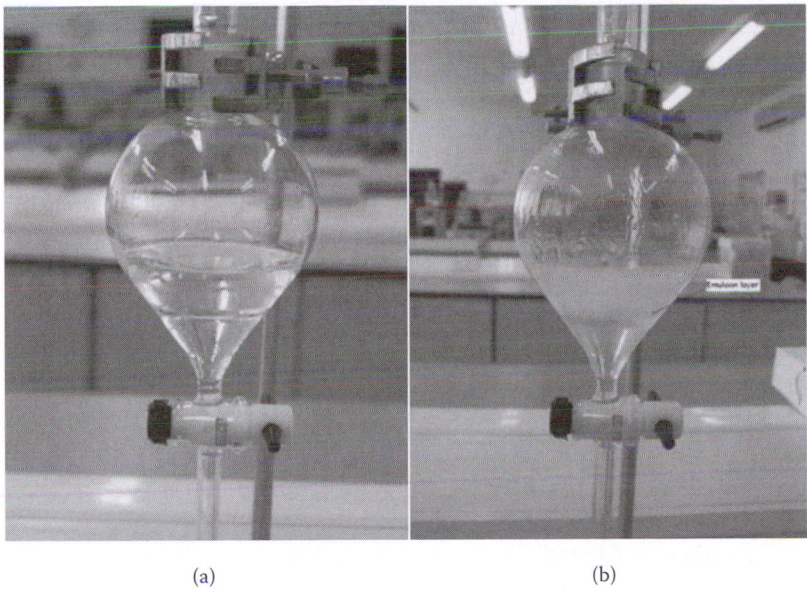

(a)                                    (b)

**Figure 3.8 (See colour insert following page 142.)** Liquid extraction. (a) Two distinct, separate layers; the lower (colourless) layer is dichloromethane and the top layer is the aqueous layer. (b) The formation of an emulsion between the two layers.

Under the experimental parameters for a liquid–liquid extraction, having two immiscible liquids, which may be considered as two separate phases (a and b), the mole fraction ratio of the solute between the phases is constant and is independent of its quantity. This law is expressed by the following equation:

$$K_p = \frac{[A]_a}{[B]_b} = \frac{\text{concentration of solute in solvent 'a'}}{\text{concentration of solute in solvent 'b'}}$$

where $K_p$ = distribution/partition coefficient (a constant).

This law is a phase of Gibbs's phase rule:

$$P + F = C + 2$$

where
   $P$ = number of phases
   $F$ = number of degrees of freedom
   $C$ = number of components

In LLE, we largely have an organic and an aqueous layer and, if we have one solute, then the number of degrees of freedom is

$$F = C + 2 - P$$

$$F = 3 + 2 - 2$$

$$F = 3$$

For further information on Gibbs's phase rule, see *Elements of Physical Chemistry* in the 'Further Reading' section at the end of this chapter.

According to the Nernst distribution law,

$$K_D = \frac{[X_2]}{[X_1]}$$

where

$K_D$ = distribution coefficient (or partition coefficient, $K_p$)
$[X_1]$ = concentration of the solute in phase 1 at equilibrium
$[X_2]$ = concentration of the solute in phase 2 at equilibrium

As previously mentioned, this law is independent of the quantity of solute present in either of the two phases. This form of the equation, however, does not take into consideration the *activity coefficient* for the solutes in the organic and aqueous phases. By taking this activity coefficient into account, we use the distribution ratio (*D*) to account for the total concentration of solute in the two phases:

$$D = \frac{\text{total concentration of solute in organic phase}}{\text{total concentration of solute in aqueous phase}} = \frac{C_2}{C_1}$$

Under idealised conditions, $K_D$ would be equal to *D*, but under experimental conditions, this does not stand. Therefore, it is more practical to use the extraction percentage (*E*), which can be related to the distribution ratio,

$$E = \frac{100.D}{D + \left(\dfrac{V_w}{V_o}\right)}$$

where

$V_w$ = volume of the aqueous phase
$V_o$ = volume of the organic phase

Here, $V_w/V_o$ is known as the phase ratio.

It is possible, with some modification of the previous equation, to determine the extraction percentage when repeated extractions are carried out because it is rare that a single volume of solvent will fully extract the solutes in the aqueous layer:

$$E = \left\{ 1 - \left[ \frac{1}{1 + D\left(\dfrac{V_w}{V_o}\right)} \right]^n \right\} \times 100$$

where $n$ = number of repeats of extractions carried out.

In toxicology, body fluids such as blood and urine can be directly extracted using liquid–liquid extraction. This is because these samples can be easily partitioned with an organic solvent without protein precipitation (however, in some laboratories, protein precipitation will be carried out on blood samples prior to LLE) but after pH adjustment. The solvent that is used in the extraction will be chosen based on its polarity and the solubility of the substances extracted within that solvent.

Another factor that is significant in LLE is the solubility that the solvent has with water. It is important that the liquids are immiscible in order to provide two layers between which the analytes can be partitioned; however, reagents can be added to produce to a distinct layer for the separation to occur. The ability of the solvent to accept and/or donate hydrogens should also be considered as the ability of the solvent to hydrogen bond with the analytes can aid or impede extractions. Table 3.7 shows typical extraction solvents with their solubility in water and their ability to hydrogen bond.

In toxicology laboratories, drug substances are grouped together for extraction and will be found in the acidic or basic layers. However, neutral compounds can be found in both the acidic and basic layers, although they do tend to predominate in the acidic layer.

**Table 3.7   Typical Extraction Solvents**

| Solvent | Solubility of Water in Solvent (g L⁻¹) | H-acceptor | H-donor | Density (g mL⁻¹) | Boiling Point (°C) |
|---|---|---|---|---|---|
| | | H-bonds | | | |
| Acetonitrile | Miscible | Yes | Yes | 0.786 | 82 |
| Acetone | Miscible | No | Yes | 0.79 | 56 |
| Ethyl acetate | 29.1 | No | Yes | 0.897 | 77 |
| Dichloromethane | 13 | No | No | 1.326 | 40 |
| Chloroform | 0.67 | No | No | 1.49 | 61 |

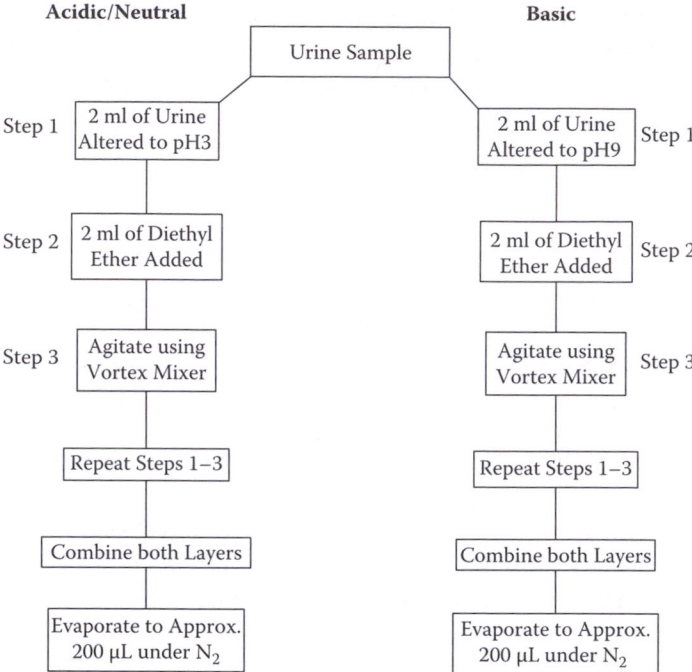

**Figure 3.9** Example of a liquid–liquid extraction protocol.

Strong acids have pKa values between 1 and 5 and weak acids have pKa values between 5 and 9; neutral compounds have neither an acidic nor a basic functional group and therefore can be extracted over the whole range of pH values. Figure 3.9 is a flow diagram of an example of a liquid–liquid extraction of a urine sample for both acidic/neutral compounds and basic compounds. In the example given in Figure 3.9, diethyl ether was used as the extraction solvent; practically speaking, it is best to use the least polar solvent that can be used to extract the solute effectively. If a solvent with too much 'extracting power' is used, this can reduce the selectivity of the method of extraction by promoting the extraction of not only the solutes of interest, but also other interfering compounds.

*Solid phase extraction* is a technique that has found increased use over the past few years in toxicology and drug labs. This is due to the fact that this technique appears to overcome some of the usual problems encountered with liquid–liquid extraction, such as emulsion formation, inclusion of impurities, and the sometimes low recovery of analytes of interest and large amounts of solvents used if carrying out many extractions.

Solid phase extraction separations are carried out by passing mixtures through an SPE cartridge, similar to the ones shown in Figure 3.10. SPE

**Figure 3.10 (See colour insert following page 142.)** Solid phase extraction cartridges of varying size.

cartridges are available in varying sizes and packing materials, depending on the sample volume and the solutes required to be extracted. Like HPLC, SPE can perform in reverse phase, normal phase, and ion exchange conditions.

### Reverse Phase SPE

With solid phase extractions in a reverse phase mode, the packing material (or stationary phase) is composed of a silica backbone that has been chemically modified with alkyl ($-CH_2-$) or aryl (any aromatic ring structure) functional groups. This packing material is nonpolar and the analytes in the mixture being extracted will also be nonpolar (with the mobile phase/sample matrix being polar). The retention and subsequent separation occur due to the analytes of interest being adsorbed by attractive forces between hydrogen and carbon bonds of the functional groups on the packing material and the analytes.

These attractive forces are known as Van der Waals forces and are weak forces of attraction (see Chapter 2 for further explanation), thus allowing the analytes adsorbed to be removed from the packing material by using a nonpolar solvent. As with reverse phase silica columns in chromatography (see Chapter 4), unreacted silanols are also present on the surface of the silica that can act as secondary reaction sites and thus can interact with polar contaminants (or polar analytes) within the mixture. As in solid phase packing material for liquid chromatography, packing material other than silica bonded phases are available and include cross-linked polymers such as styrene-divinylbenzene (see Chapter 4).

## Normal Phase SPE

In normal phase solid phase extraction, the stationary phase is composed of a modified silica polar material. The analytes of interest will be polar, in order to have an affinity for the packing material, and the matrix should be nonpolar, in order for the extraction to take place.

In comparison to reverse phase SPE, the separations in normal phase SPE occur due to the polar functional groups of the packing material and the polar functional groups of the analytes of interest interacting. The interactions that occur are based on more than one intermolecular force and include hydrogen bonding, dipole–dipole interactions, and $\pi$–$\pi$ interactions, as well as others.

In order for the separation to take place, a more polar solvent (than the original sample matrix) will be used to effect the desorption of the analyte molecules from the packing material. Examples of packing material include silica bonded with cyano, amine, and diol groups, as with the stationary phase of columns in normal phase chromatography (see Chapter 4 for further explanation).

## Ion Exchange SPE

Ion exchange SPE cartridges will be anion exchange or cation exchange cartridges. These cartridges function on the basis of electrostatic interactions between the charged groups of the solutes in the sample and the charged groups on the surface of the packing material. Strong and weak cation exchange packing material and strong and weak anion exchange packing materials are available. The terms 'strong' and 'weak' refer to the extent to which the ionisation of the packing material varies with pH and not the degree of strength of binding to analyte molecules. Packing materials— again, such as with those used in ion exchange chromatography stationary phases—are sulphonic acid groups (strong cation exchanger), carboxylic acid groups (weak cation exchanger), quaternary amine groups (strong anion exchanger), and tertiary amine groups (weak anion exchanger), although others are available. See Chapter 4 for further information.

## SPE Procedures for Extraction

Practically, when using solid phase extraction cartridges, it is first necessary to condition the packing material. This is done by rinsing the cartridge with a solvent, which will be chosen depending upon the type of cartridge being used (whether normal phase, reverse phase, or ion exchange). When this solvent has passed through the column, the sample solution will be added to the cartridge. The cartridge will be used in one of two ways:

The analytes of interest will adsorb onto the packing material and the
rest of the matrix solution with other components will pass through
unretained OR

The matrix solution with other components will adsorb onto the packing material and the analytes of interest will pass through the column unretained.

Either way, the cartridge will be washed to ensure that all of the analytes
of interest have passed through or that all of the other components have
passed through.

If the analytes of interest are adhered to the packing material, a solvent
will be chosen that will effect their desorption in order to be collected and
further analysed. The most commonly used method is to have the analytes
of interest adsorbed onto the packing material, thus allowing the other components in the matrix to pass through unretained rather than the opposite
scenario. This is because retaining the analytes generally results in a more
efficient extraction.

## Sample Dilution

Once the compound of interest has been isolated from the sample matrix, it
is often necessary to dilute the sample in order to analyse it within the validated assay range (see Chapter 8). A test injection may have to be performed
in order to assess the sample concentration in the first instance. Once it has
been established that the sample does require dilution, this can easily be performed using calibrated glass (larger volumes) or variable volume pipettes
(smaller volumes).

### *Glass Volumetric Pipettes*

Draw up the desired volume into the pipette and transfer this to a suitable
volumetric flask. Dilute to volume with solvent and invert to mix. Further
dilutions can be performed in the same manner.

### HYPOTHETICAL EXAMPLE 3.3

You have been asked to dilute a drug solution 10 times, using mobile phase as the
solvent. You have a starting volume of 50 mL (solution A). Using a 5 mL volumetric
pipette, draw up 5 mL of solution A and dispense this into a 50 mL volumetric
flask. Dilute to volume with mobile phase, stopper, and invert to mix. The sample
solution has been diluted 10 times. This is often referred to as a 1 in 10 or 1 to 10
dilution. It should not be confused with a 1:10 dilution, which can be interpreted
as meaning 1 mL of solution plus 10 mL of diluent, which would be a 1 in 11 or
1 to 11 dilution. It is always wise to check a final dilution with a more experienced
operator, such as a university laboratory technician or a more senior member of
staff, just to be sure that you have correctly understood the instruction.

## *Variable Volume Pipettes*

Set the pipette to the desired volume and draw this up into the pipette. Dispense the solution into a suitable receiving vessel; this might be an HPLC vial or another small glass or plastic vial. Using a similar type of pipette, transfer the diluents into the same vessel. Shake or invert to mix.

---

### HYPOTHETICAL EXAMPLE 3.4

You have been asked to dilute a liver extract 10 times using mobile phase as the solvent. Following an extraction procedure, you have a volume of 200 μL (solution A). Using a variable volume automatic pipette, draw up 100 μL of solution A and dispense this into a suitable glass or plastic vial. Using a second variable volume automatic pipette, transfer 900 μL of mobile phase into the same vial. Stopper the vial and shake or invert to mix. The sample solution has been diluted 10 times. This is often referred to as a 1 in 10 or 1 to 10 dilution.

---

## Reference Standard Preparation and Dilution

Reference standards very rarely require any sophisticated pretreatment before they are ready for use, unless they are to be derivatised in some way. In most cases, the reference standard is weighed and quantitatively transferred into a suitable volumetric flask with the aid of the solvent. Any dilutions that are required are performed in the same way as the sample descriptions given previously, again depending on the sample size.

One major difference with standard solutions is the requirement to construct what is known as a standard or calibration curve. This is a series of standards of known, but different, concentrations that are used to calculate the concentration of the unknown sample solutions. The calibration curve is a linear regression plot of peak area versus concentration. The closer the regression line is to a straight line ($R^2 = 1$), the more accurate is the calculation of the unknown samples.

The individual standards used to construct the curve can be diluted in a number of ways. One way is to perform what are called 'serial dilutions'. This results in a series of solutions covering a wide range of concentrations. The second way in which to perform dilutions is to prepare a stock solution and further dilute aliquots from this solution to the desired levels. This does not have the disadvantage associated with serial dilutions and allows for a much tighter or smaller range of concentrations.

### *Serial Dilutions*

Serial dilutions are performed by preparing a 'stock' solution of known concentration at a level equal to or in excess of the most concentrated sample requirement. The stock solution is then diluted to produce a secondary solution, which is in turn diluted and so on until all of the required concentrations

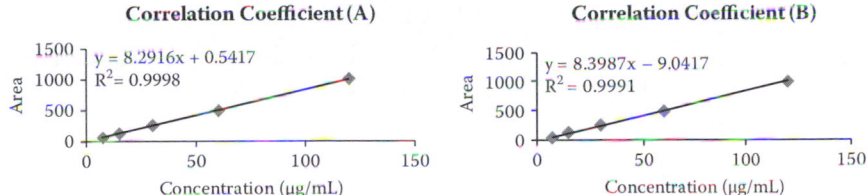

**Figure 3.11** Demonstration of error in correlation coefficient ($R^2$).

have been achieved. The dilution factor at each dilution step remains constant. When there is a 10-fold dilution at each stage, the serial dilution is known as a logarithmic dilution.

It should be noted, however, that the use of a serial dilutions approach will depend on the range that has been established for the method during the validation phase. This is because serial dilutions often result in very broad ranges of concentration. The disadvantage of this is that the correlation coefficient ($R^2$) can remain relatively unaltered, with quite substantial errors in measurement—particularly at the low concentration end of the plot. In Figure 3.11, we can see that a 40% error in the measurement at 7.5 µg/mL will still give an acceptable $R^2$ (correlation coefficient B) when the limit has been set at $R^2 \geq 0.999$.

### HYPOTHETICAL EXAMPLE 3.5

Prepare a series of reference standards containing the following concentrations of drug substance: 120, 60, 30, 15, and 7.5 µg/mL. You have been provided with a stock solution containing 240 µg/mL (S):

10 mL S diluted to 20 mL with solvent = 120 µg/mL (A)
10 mL A diluted to 20 mL with solvent = 60 µg/mL (B)
10 mL B diluted to 20 mL with solvent = 30 µg/mL (C)
10 mL C diluted to 20 mL with solvent = 15 µg/mL (D)
10 mL D diluted to 20 mL with solvent = 7.5 µg/mL (E)

In this case, we have a dilution factor of 2 at each stage.

## Dilutions from Stock

Dilutions from stock are prepared from the stock solution by removing aliquots and diluting these to the desired level. The dilution factor will change with each newly prepared solution. This does not have the disadvantage associated with serial dilutions and allows for a much tighter or smaller range of concentrations. The disadvantage here is that it can be difficult to dilute the stock solution to exactly the required concentration. The standard curve can be adjusted to reflect this and this should be validated in the linearity section of the method validation and recorded in the standard operating procedure.

**HYPOTHETICAL EXAMPLE 3.6**

Prepare a series of reference standards containing the following concentrations of drug substance: 150, 120, and 75 µg/mL. You have been provided with a stock solution containing 300 µg/mL (S):

10 mL S diluted to 20 mL with solvent = 150 µg/mL
20 mL S diluted to 50 mL with solvent = 120 µg/mL
25 mL S diluted to 100 mL with solvent = 75 µg/mL

## Sample Introduction—Autosamplers

Injecting the sample into the HPLC system can be done manually using a syringe or via an autosampler. In either case, the mechanism for injection is via a Rheodyne® valve (see Figure 3.12). The Rheodyne system operates using a six-way valve with high-pressure switching taking place between a rotor seal and a stator face. Most HPLC systems come equipped with an auto-sampler, which offers greater flexibility compared to the manual systems.

### Manual System

With a manual system, the sample is injected via a syringe through the nee-dle port into a fixed volume loop. The handle is then turned to the inject position and the sample is swept out of the loop, with the aid of the mobile phase, and onto the column. The syringe is then removed and the handle is returned to the load position. Loop sizes can vary from 2 µL to 5 mL. Some

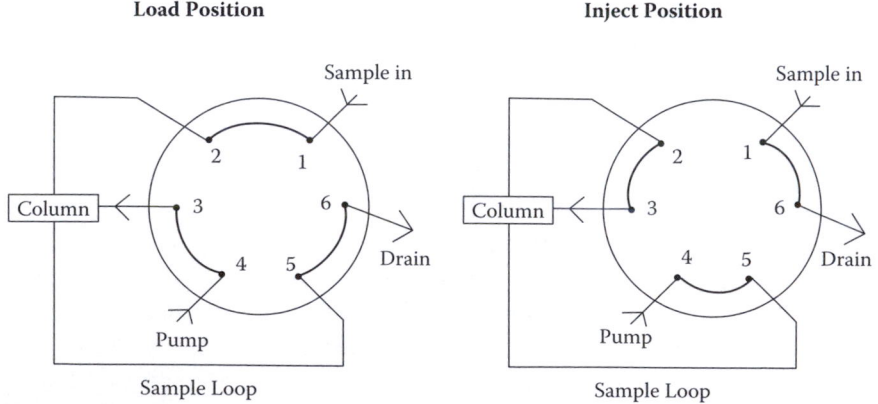

**Figure 3.12** Rheodyne® valve.

manual systems allow operation in both complete- and partial-fill modes while maintaining accurate delivery of sample volume. There is the risk of error in relation to the volume delivered when using a partial-fill method and only those systems designed for that purpose should be used.

The major disadvantage of a manual system is that the operator must be present to make the injection. This can be time consuming and also means that the system cannot be used when there are no operators present, such as overnight and during lunch and dinner breaks. In the forensic environment, it is often necessary to run many samples at a time; therefore, the manual systems have been replaced by automated systems capable of handling up to 100 samples at any one time.

## Automated System (Autosampler)

Automated systems offer greater flexibility in relation to operator input and can be left unattended for long periods of time. The downside of this is that the critical information required to run the system correctly must be manually entered. If errors occur, they can be difficult to spot and, as a consequence, the whole, or part of the sequence of injections may need to be repeated. Many laboratories now include a checking system into their quality procedures to try to prevent this from happening.

Autosamplers work using the same switching valve principles as the manual system. A metered syringe draws a sample through a needle into a loop, the rotor turns against the stator face, and the sample is flushed onto the column. Carryover of the sample from the needle and from the switching valve can be an issue. Many manufacturers now operate a continuous flow system whereby mobile phase is continually pumped through the loop and sample needle when the system is operational.

During the loop filling stage, the mobile phase is diverted directly through the column. The volume of the injection can be varied, depending on the loop size, using the syringe function. A needle wash function is also available on some systems; this ensures that the outside of the needle remains 'clean' and sample carryover is greatly reduced. If viscous samples are to be injected, then the sample draw speed into the needle can be altered on some systems to account for this. If a viscous sample is drawn too quickly into a syringe needle, air is often introduced and this will interfere with the volume injected and also with the chromatography (see Chapter 10).

Autosamplers can be temperature controlled. This is desirable when samples need to be kept refrigerated or when there is a need to heat the sample prior to injection.

**KEY POINTS SUMMARY**

*Mobile phase:* always use high-quality reagents and filter and degas before use.

*Mixing of mobile phase:* this can be performed manually or by using binary or quaternary HPLC instrumentation. When mixing mobile phase manually, always ensure that individual components are dispensed separately and then mixed. Do not readjust the final solution in terms of volume.

*Buffer pH:* when working with buffer, always ensure that you are working within ±1 pH unit of the buffer pKa to ensure maximum buffering capacity. Concentration is not of vital importance, assuming that sufficient buffer is present. To avoid buffer salts coming out of solution, it is advisable to work at as low a concentration as possible and to use low levels of organic modifier in the mobile phase.

*pKa:* optimum resolution in RP-HPLC is obtained by ensuring that the analyte is non-ionised. This means operating at a pH above the pKa for basic compounds and below the pKa for acidic compounds. It is not always possible to work at such extremes of pH, so for highly basic and acidic compounds (which might adhere to the column), ion exchange chromatography may be required.

*Sample extraction:* it may be necessary to separate out the compounds of interest from the sample matrix by using SPE or LLE.

*Sample dilution:* dilutions are normally performed either as serial dilutions or as dilutions from a preprepared stock solution. Serial dilutions give a much wider concentration range and care should be taken to ensure that the concentrations fall within the validated range of the method.

*Sample injection:* manual injection and autoinjection are available; however, more errors might be expected with a manual injection system.

**QUESTIONS**

1. Give one reason why it is necessary to filter mobile phase solutions before use on an HPLC system.
2. Give one reason why it is necessary to degas mobile phase solutions before use on an HPLC system.
3. You have been asked to prepare a mobile phase containing methanol, acetonitrile, and aqueous 10 m$M$ phosphate buffer in the proportions 30:10:60. Write down the shorthand version of this mobile phase.

4. You have been asked to prepare a mobile phase containing a small amount of triethylamine. The validated wavelength for the method is 220 nm. What issues might arise from this combination?

5. You are running an HPLC method for the first time and have encountered several unknown small peaks in the chromatograms. You review the method development data, but there is no mention of impurity peaks. Give one example of a cause for these small peaks.

6. Describe what is meant by the terms *acid* and *base*.

7. List the following in order of acidity (most acidic first): ammonia, hydrochloric acid, nitric acid, and sulphuric acid.

8. Ketamine is an anaesthetic drug used in both human and veterinary medicine. Because it is commonly abused due to its hallucinogenic effects, it is a class C controlled drug under the Misuse of Drugs Act. Using the information that follows, answer the following questions:

   a.  If you buffered your mobile phase at pH 3.5, would ketamine be in the ionised or non-ionised form?

Ketamine pKa = 7.5

   b.  Give an example of a suitable buffer that could be used to prepare the preceding mobile phase.

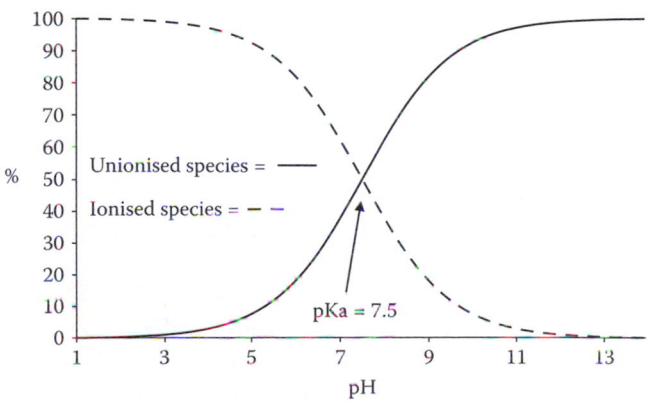

   c.  If you increased the pH of the mobile phase to pH 11, what might you expect to happen to the retention time (e.g., would you expect the retention time to increase or decrease on a reversed phase HPLC column, all other parameters remaining the same?)? Explain your reasoning.

9. Why is it necessary to use a buffer only within ±1 pH unit of the buffer pKa?
10. It is desirable to use as low a buffer concentration as possible in HPLC systems. Why is this the case and how can problems associated with this be rectified?
11. Given the following information, which sample preparation technique(s) would you choose and why? If you were to decide on an SPE cartridge for extraction, which one would you choose and why?

| Morphine | Codeine | Morphine-6-glucuronide |

## Further Reading

### Books/Online Publications

Agilent Technologies. *Effect of low pH on retention time* (www.home.agilent.com).

Atkins, P., and J. DePaula. 2005. *Elements of physical chemistry,* 4th ed. Oxford, England: Oxford University Press.

Bidlingmeyer, B. A. 1992. *Practical HPLC methodology and applications.* New York: Wiley-Intersciemce.

*Clinical/forensic products and applications for GC and HPLC,* Restek technical literature. 2006/2007 (www.restek.com).

Heyrman, A. N., and R. A. Henry. Importance of controlling mobile phase pH in reversed phase HPLC. Keystone technical bulletin TB 99-06 (www.hplcsupply.com).

Kazakevich, Y., and R. LoBrutto. 2007. *HPLC for pharmaceutical scientists.* New York: John Wiley & Sons.

Kenkel, J. 2002. *Analytical chemistry for technicians,* 3rd ed. Boca Raton, FL: CRC Press.

Laboratory and Scientific Section, United Nations Office on Drugs and Crime in Cooperation with the Drugs Working Group of the European Network of Forensic Science Institutes. 2009. *Guidelines on representative drug sampling* (www.enfsi.com).

Laserchrom. *Understanding pH buffers. Which one to use and at what concentration* (www.laserchrom.com).

*Measurement uncertainty arising from sampling—A guide to methods and approaches,* Eurachem/CITAC guide. 2007. Produced jointly with EUROLAB, Nordtest, and the UK RSC Analytical Methods Committee.

Pawliszyn, J. 2002. *Sampling and sample preparation for field and laboratory: Fundamentals and new directions in sample preparation.* New York: Elsevier.

Smith, F. P., J. A. Siegel, and S. A. Athanaselis. 2005. *Handbook of forensic drug analysis.* San Diego, CA: Academic Press.

Standard Committee for Quality and Competence, European Network of Forensic Science Institutes. 2007. Guidance for best practice sampling in forensic science—QCC-BPS-001.

## Journals

Baiocchi, C., M. C. Brussino, E. Pramauro, A. B. Prevot, L. Palmisano, and G. Marci. 2002. Characterisation of methyl orange and its photocatalytic degradation products by HPLC/UV-VIS diode array and atmospheric pressure ionisation quadrupole ion trap mass spectrometry. *International Journal of Mass Spectrometry* 214: 247–256.

Bates, J. W., and J. A. Lambert. 1991. Use of the hypergeometric distribution for sampling in forensic glass comparison. *Journal of the Forensic Science Society* 31 (4): 449–455.

Boonkerd, S., M. R. Detaevernier, J. Vindevogle, and Y. Michotte. 1996. Migration behaviour of benzodiazepines in micellar electrokinetic chromatography. *Journal of Chromatography A* 756: 279–286.

Coopman, V., M. De Leeuw, J. Cordonnier, and W. Jacobs. 2009. Suicidal death after injection of a castor bean extract (*Ricinus communis* L.). *Forensic Science International* 189: e13–e20.

Lopes Marques, R. M., P. J. Schoenmakers, C. B. Lucasius, and L. Buyden. 1993. Modelling chromatographic behaviour as a function of pH and solvent composition in RPLC. *Chromatographia* 36 (1): 83–95.

Maralikova, B., and W. Weinmann. 2004. Confirmatory analysis for drugs of abuse in plasma and urine by high-performance liquid chromatography—Tandem mass spectrometry with respect to criteria for compound identification. *Journal of Chromatography B* 811: 21–30.

Moeller, M., S. Steinmeyer, and T. Kraemer. 1998. Determination of drugs of abuse in blood. *Journal of Chromatography B* 713: 91–109.

Nishida, M., A. Namera, M. Yashiki, and K. Kimura. 2004. Miniaturised sample preparation method for determination of amphetamines in urine. *Forensic Science International* 143 (2–3): 163–167.

Wang, X., J. Yu, M. Xie, Y. Yoa, and J. Han. 2008. Identification and dating of the fountain pen ink entries on documents by ion-pairing high-performance liquid chromatography. *Forensic Science International* 180: 43–49.

Wiczling, P., M. J. Markuszewski, M. Kaliszan, K. Galer, and R. Kaliszan, 2005. Combined pH/organic solvent gradient HPLC in analysis of forensic material. *Journal of Pharmaceutical and Biomedical Analysis* 37 (5): 871–875.

Yazdi, A. S., and Z. Es'haghi. 2005. Surfactant enhanced liquid-phase microextraction of basic drugs of abuse in hair combined with high performance liquid chromatography. *Journal of Chromatography A* 1094: 1–8.

# Modes of Separation

4

## Introduction

In the second chapter of this book, you were introduced to the basic theory of chromatography. Here in this chapter, we shall look more specifically at the modes of separation used in HPLC. Already, the concept of a mobile phase and a stationary phase has been established. Here, we further discuss this basic concept and explain the chemical nature of the stationary and mobile phases and how varying these, as well as considering the analytes we have, has an effect on the separations that can be carried out.

In liquid chromatography, separations will occur based on one of two principles: *partition* and *adsorption*. With partition (*liquid–liquid chromatography*), a liquid stationary phase is used. This phase will be coated onto or chemically bonded onto a finely divided inert support (the latter is more common). With all forms of liquid chromatography, as the name suggests, we have a liquid mobile phase.

For separation to occur with this type of chromatography, the liquid sample will be dispersed in the mobile phase with the analytes of interest being partitioned between the mobile phase and the stationary phases, depending upon their partition coefficients. As a reminder, the partition coefficient, $K_p$, is expressed in the following equation:

$$K_p = \frac{[A]_a}{[B]_b} = \frac{\text{concentration of solute in solvent 'a'}}{\text{concentration of solute in solvent 'b'}}$$

The difference in partition coefficient values leads to differential rates of migration of the analytes, thus affecting a separation.

In adsorption (*liquid–solid chromatography*), a solid stationary phase is used where a surface phenomenon, known as adsorption, affects the separation. Again, because we are dealing with liquid chromatography, the mobile phase is a liquid. The separation is affected by the equilibrium that is reached between the molecules that are adsorbed onto the stationary phase and those unbound in the mobile phase.

The columns (Figure 4.1) are tubular in shape; dimensions of the column are given in the following order: internal diameter × length, particle size of the packing material (e.g., 4.6 × 150 mm, 3.5 μm). These columns vary in their

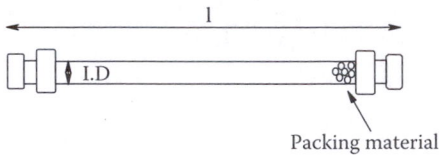

Packing material

**Figure 4.1** HPLC column.

internal diameters, lengths, and particle size (which will affect the pore size), but the choice will depend upon a number of factors, including the compounds being analysed and the flow rates required. Practically, larger particle sizes will produce lower overall pressure within the system; conversely, smaller particle sizes will increase the pressure within the system and smaller particles will generally produce higher separation efficiencies (see Chapter 6 for further explanation).

We can alter the mobile and stationary phases to suit the chemical nature of the compounds in a sample that we wish to analyse. By altering the composition of the mobile phase or by changing the polarity of the stationary phase, we change the mode in which the separation is taking place. The modes most commonly encountered and used in HPLC are

- reverse phase
- ion exchange
- normal phase

Other modes, such as chiral chromatography, also exist, but find very little, if any, application in forensic science applications. Should you wish to know more about any of these modes, please see the 'Further Reading' section at the end of this chapter.

## Reverse Phase HPLC

Reverse phase chromatography systems use a nonpolar stationary phase with a polar mobile phase. Generally, the packing materials within the reverse phase columns are composed of spherical silica beads coated with hydrophobic alkyl chains ($-CH_2-CH_2-CH_2-$) of varying length. Typical coating materials are C4, C8, and C18; the longer the chain is, the more nonpolar the *stationary phase* will be. Other column packing materials are also found in reverse phase chromatography and these can include phenyl. These columns will be used with *mobile phases* containing mixtures of aqueous and organic solvents. Table 4.1 outlines the main differences of the two most commonly used columns, C8 and C18 (C18 the most widely used), and phenyl columns.

**Table 4.1  Main Differences of the C8, C18, and Phenyl Columns**

| Phase | Functionality | Bond Type | Additional Information |
|---|---|---|---|
| C18, octadecylsilane | n-C$_{18}$ | Si–O–Si–C | • pH range 2–8[a]<br>• Most commonly used RP column packing material |
| C8, octylsilane | n-C$_{8}$ | Si–O–Si–C | • pH range 2–8[a]<br>• These phases tend to have lower resolution than ODS phases<br>• Useful when varying component polarities are present in a mixture |
| Phenyl | | Si–O–Si–C | • pH range 2–8[a]<br>• These phases tend to have lower resolution than both ODS and OS phases<br>• These columns can also be used in normal phase chromatography |

[a] These ranges can be extended from 1.5 to 10; however, special bonded phase columns must be purchased.

Because silica with C8 and C18 phases is used predominantly, our discussions of surface interactions and thus separations will be based around these packing materials. Should you wish to find out more about the chemistry of other packing materials, such as phenyl, please see the 'Further Reading' section at the end of this chapter.

The separations occur due to the interactions of the alkyl ($-CH_2-$) groups on the analytes within the mixture and the *functional groups* on the surface of the silica. These interactions are known as Van der Waals forces (see Chapter 2 for further explanation), but the mechanisms of retention of analytes are complex and it is possible that more than one mechanism is operating at the same time. For example, hydrophobic interactions may be taking place on the surface of the bonded stationary phase, which can cause partitioning effects with solutes as well as adsorption effects on unreacted silanol ($-Si-OH-$) groups.

Unreacted silanol groups are considered to be strong adsorption sites and are hydrophilic in nature. These silanol groups are present because they are supposed to serve as a support for alkyl groups (such as the C8 and C18 groups) used in bonded phases. Due to factors such as steric hindrance, silanol groups are found unreacted (as can be seen in Figure 4.2).

These silanol groups are found in three different forms: silanol groups, which are hydrogen bonded to an adjacent oxygen atom in an adjacent silanol group (called vicinal groups); reactive groups (called germinal groups); and free silanols (called isolated groups). Theses silanol groups are shown in Figure 4.3.

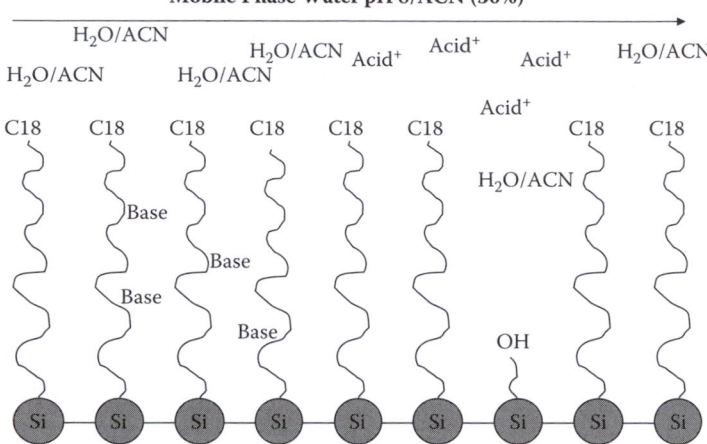

**Figure 4.2 (See colour insert following page 142.)** Example of RP-MPLC separation.

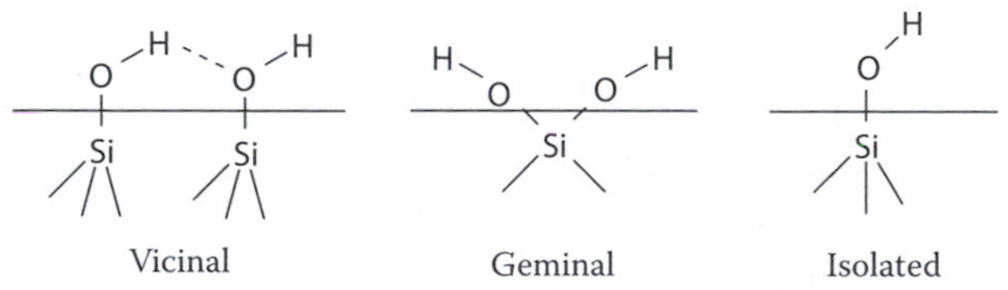

**Figure 4.3** Forms of silanol groups.

These silanol groups are weakly acidic and their presence can cause difficulty when analysing basic compounds. Chromatographic silicas usually contain between 0.1 and 0.3% metallic impurities. The presence of these metals causes silanols to become highly acidic and thus can influence greatly the chromatographic properties; this can lead to poor resolution of peaks.

In order to overcome the problems in relation to unreacted silanol groups, end-capping was introduced. This is a process in which the packing material goes through a secondary bonding phase to eliminate unreacted silanols, but it also prevents dissolution of underlying silica during chromatography.

In reverse phase chromatography, the mobile phase is more polar than the stationary phase and can be varied by changing the mobile phase composition. As mentioned before, the mobile phase used tends to be a combination of aqueous and organic solvents; by increasing the aqueous component, we can produce a more polar mobile phase and by increasing the organic component we can render the mobile phase less polar (or more nonpolar). In the reverse phase system, nonpolar solutes will have a greater affinity for the stationary phase, with the more polar components of the mixture having a greater affinity for the mobile phase.

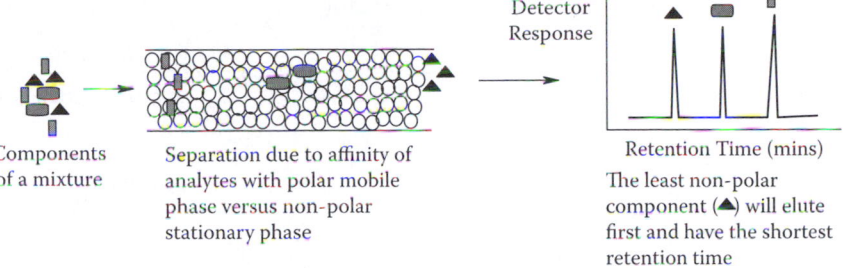

| Components of a mixture | Separation due to affinity of analytes with polar mobile phase versus non-polar stationary phase | Retention Time (mins)<br><br>The least non-polar component (▲) will elute first and have the shortest retention time |

**Figure 4.4** (See colour insert following page 142.) Separation in RP-chromatography.

Figure 4.4 shows the separation of a mixture of compounds of varying degrees of polarity. The least nonpolar compound will elute first and thus have the shortest retention time, whereas the most nonpolar compound will be retained on the column longer and thus will have the longest retention time.

## Ion Exchange

Ion exchange chromatography utilises the charge immobilised on the ion exchange stationary phase (resin) and the charge on the sample. Ion exchange chromatography can be subdivided into cation and anion exchange, with the analytes being those that are ionic, or easily ionisable. Resin materials are composed of cross-linked polydivinylbenzene (the higher the degree of cross-linking, the harder is the resin—rendering it less susceptible to swelling, although this property can reduce the ion exchange capacity of the resin), sulphonic acid groups, carboxylic acids, and amines. The retention mechanism in ion exchange chromatography is predominantly based on electrostatic attraction of the analyte species and the charged functional group of the packing material (bonded silica). Table 4.2 outlines the differences in ion exchange packing materials.

Unlike in other forms of chromatography, where 'like dissolves like' is the basic theory used, in ion chromatography we use the analogy of 'opposites attract'. Practically speaking, for cation and anion exchange systems, the sample is loaded onto the column and binds to the resin depending on charge (of the analytes and the resin), where an initial buffer (of low ionic strength) is used to allow the system to equilibrate. The bound buffer molecules (counter-ions) are displaced by the analyte molecules. Then the counter-ions are washed from the column by using a second buffer, which will be increased steadily (gradient) in ionic strength. During this time, the analyte molecules will be desorbed with the order of elution of the analytes determined by the external charge on each of the compounds in the sample mixture.

**Table 4.2    Differences in Ion Exchange Packing Materials**

| Phase | Functionality | Bond Type | Additional Information |
|---|---|---|---|
| Benzene sulphonic acid | $-CH_2CH_2$ $-SO_3-H^+$ | Si–O–Si–C | • Strong cation exchange sorbent<br>• Divalent ions more strongly retained than monovalent<br>• Can also separate on the basis of nonpolar interaction (due to presence of benzene) |
| Propyl sulfonic acid | $-CH_2.CH_2.CH_2$ | Si–O–Si–C | • Weak anion exchange sorbent<br>• Can be used with weaker cations<br>• Can be used with biological and other aqueous samples |
| Bonded amine | $-CH_2.CH_2.CH_2.NH_2$ | Si–O–Si–C | • Weak cation exchange resin<br>• Protonates below pH 9.8<br>• Phase is less stable than cyano or diol phases in ion exchange chromatography |
| Propyl, ethylene diamine | $-CH_2.CH_2.CH_2.NH. CH_2.$ $CH_2.NH_2$ | Si–O–Si–C | • Weak anion exchange resin<br>• Can be used with biological and other aqueous samples |

## Cation Exchange

In cation exchange, the positive charge on the sample binds to the negative charge on the resin. Consider the following exchange of two cations, $A^+$ and $B^+$, between the solution containing the cations and the ion exchange resin ($R^-$):

$$AR+B^+ \rightleftharpoons BR+A^+$$

The equilibrium constant for this reaction is shown in the following equation:

$$K_a = \frac{[A^+][BR]}{[AR][B^+]}$$

Using this value ($K_a$), we can determine the affinity that the cations ($A^+$ and $B^+$) will have for the oppositely charged sites on the stationary phase resin. When $K_a = 1$, the system will not be able to discriminate effectively between the two ions and thus a separation will not take place.

Typically, two types of ion exchange resins are used in cation exchange mode: a strong cation exchanger (e.g., sulphonic acid group, $SO_3^-$) and a weak cation exchanger (e.g., carboxylic acid group, $CO_2^-$). Generally, cation exchange is carried out using buffers over a pH range in a gradient from 4 to 7. The separation takes place due to the ionised sites on the resin and the attraction that analytes have for these sites.

### Anion Exchange

In anion exchange, the negative charge on the sample binds to the positive charge on the resin. Consider the exchange of two anions, $X^-$ and $Y^-$, again between the solution and the resin ($R^+$),

$$XR + Y^- \rightleftharpoons YR + X^-$$

The equilibrium constant, $K_a$, is expressed as

$$K_a = \frac{[X^-][YR]}{[XR][Y^-]}$$

Typically, two types of ion exchange resins are used in anion exchange mode: a strong anion exchanger (e.g., quaternary amine group, $-NR_3^+$) and a weak anion exchanger (e.g., tertiary amine group, $NR_3$). Generally, anion exchange is carried out over a range of pH from 7 to 10 and, like cation exchange, this is done in a buffer over a gradient.

## Normal Phase HPLC

Normal phase chromatography systems use a polar stationary phase with a nonpolar mobile phase. Generally, the packing materials within the normal phase columns are composed of unmodified silica spherical beads (cyano, amine, or diol packing materials can also be used) with the mobile phase consisting of nonpolar organic solvents such as ethanol, chloroform, propanol, or hexane. Table 4.3 outlines the main differences between typical normal phase packing materials.

The separations occur due to the most polar compounds within the mixture having a higher affinity for the stationary phase and the least polar compounds having a higher affinity for the mobile phase (as is shown in Figure 4.5).

In this mode of separation, in comparison with reverse phase, the most nonpolar compounds will elute first and have the shortest retention time, whereas the least nonpolar compounds will elute later (see Figure 4.6).

**Table 4.3    Main Differences in Typical Normal Phase Packing Materials**

| Phase | Functionality | Bond Type | Additional Information |
|-------|---------------|-----------|------------------------|
| Bonded diol | OH  OH | Si–O–Si–C | • pH 3–8<br>• H-bonding similar to levels found in unbonded silica |
| Pure silica | Si–OH | SiO$_2$ | • pH 2–6<br>• Can be unpredictable because surface interactions can take place |
| Phenyl | | Si–O–Si–C | • pH 2–8[a]<br>• These columns can also be used in reverse phase chromatography |
| Bonded amine | –CH$_2$.CH$_2$.CH$_2$.NH$_2$ | Si–O–Si–C | • Protonates below pH 9.8 |
| Bonded nitrile | –CH$_2$.CH$_2$.CH$_2$.CN | Si–O–Si–C | • Less sensitive to mobile phase impurities than silica |

[a]  This range can be extended from 1.5 to 10; however, special bonded phase columns must be purchased.

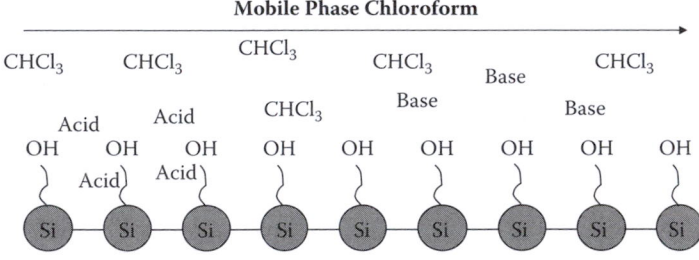

**Figure 4.5 (See colour insert following page 142.)** Examples of NP-HPLC separation.

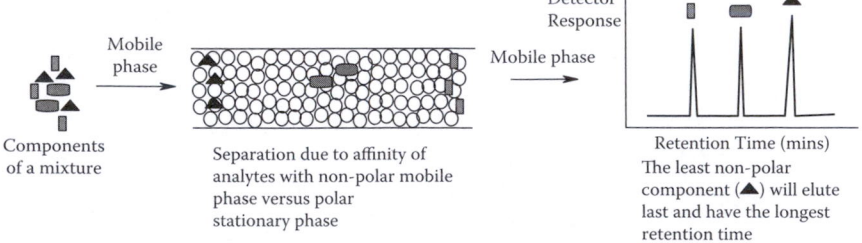

**Figure 4.6 (See colour insert following page 142.)** Separation in NP-chromatography.

---

**KEY POINT SUMMARY**

- *Reverse phase chromatography:* the mobile phase is more polar than the stationary phase.
- *Ion exchange chromatography:* cation and anion exchange is possible where the cation exchange has a negatively charged resin to attract the cations and anion exchange has a positively charged resin to attract the anions.
- *Normal phase chromatography:* the stationary phase is more polar than the mobile phase.

---

**QUESTIONS**

1. Explain the difference between adsorption and partition chromatography.
2. Phenyl and C18 are two packing materials that can be used in reverse phase chromatography; explain the differences in physical characteristics.
3. You are required to separate amitriptyline and nortriptyline using HPLC (chemical structures shown here).

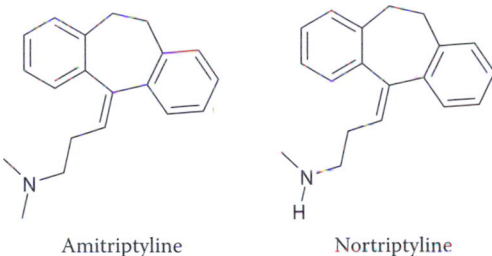

Amitriptyline               Nortriptyline

Which mode of separation would you choose and why?
4. Why are unreacted silanol groups sites for strong adsorption?

# Further Reading

Cazes, J., and R. Scott. 2002. *Chromatography theory.* Boca Raton, FL: CRC Press.
Heftmann, E. 2004. *Chromatography: Fundamentals and techniques.* New York: Elsevier.
Jönsson, J. A. 1987. *Chromatographic theory and basic principles.* Boca Raton, FL: CRC Press.

# Detection Systems

5

## Introduction

Detection is evidently an important step in chromatographic analyses because, without it, we would not have any results. As with the modes of separation, the types of columns that we can use, and the compositions of mobile phases that we can use, detectors are no different. A number of different detectors can be used but, ultimately, the compounds being analysed will determine this. This is because the detector must be able to detect the compounds eluting from the chromatographic column.

The most common method of detection in HPLC exploits the ultraviolet and visible regions of the electromagnetic spectrum (EM; see Figure 5.1) in order to detect the analytes of interest. The detectors employed that utilise these regions of the EM spectrum are the ultraviolet (UV) and diode array detectors. However, other, more specific detectors can be used for specialised applications, such as conductivity and refractive index (not discussed here), and with the introduction of hyphenated techniques, mass spectrometry has become a widely used detector method.

## Theory

The electromagnetic spectrum, shown in Figure 5.1, is composed of electric and magnetic waves that travel at a constant velocity of $2.99 \times 10^{-8}$ ms$^{-1}$ (which is the velocity of light in a vacuum). The electric and magnetic waves can vary in their wavelength ($\lambda$) represented in Figure 5.2. Wavelengths are given in nanometres (nm) and 1 nm $\equiv 1 \times 10^{-9}$ m. This is represented in the following equation:

$$c = v\lambda$$

where

$c$ = velocity of light in a vacuum
$v$ = frequency (the number of peaks passing a given point in 1 second)
$\lambda$ = wavelength

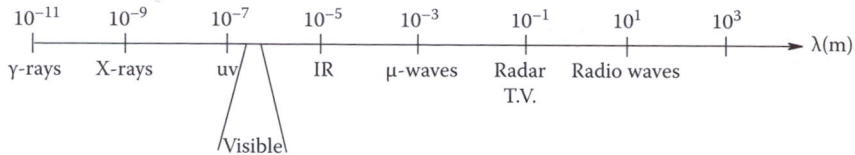

**Figure 5.1** Electromagnetic spectrum.

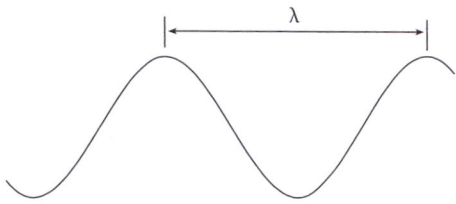

**Figure 5.2** Representation of wavelength (λ).

Here, $c$ is a constant and $v$ and $\lambda$ are inversely proportional. We are concerned with only the ultraviolet (10–380 nm) and visible (380–780 nm) regions of the electromagnetic spectrum when considering UV/Vis and diode array detectors.

In order to understand fully the practical aspects of absorption theory, it is necessary to digress briefly into quantum theory. For this explanation, let us consider that radiation takes the form of a stream of particles of 'packets' of energy, called photons (instead of the waves as previously mentioned). Each atom and molecule exists in a number of energy levels, or states, and a change of level will require the absorption (or emission) of an integral number of photons (or unit energy called a quantum, in quantum theory).

The energy of a photon adsorbed (or emitted) during one of these transitions from one molecular energy level to another is given in the following equation:

$$E = hv$$

where
$E$ = energy
$h$ = Planck's constant ($6.626 \times 10^{-34}$ Js)
$v$ = frequency of the photon

We have already seen that $c = v\lambda$; these two equations can be combined and are given in the following:

$$E = \frac{hc}{\lambda}$$

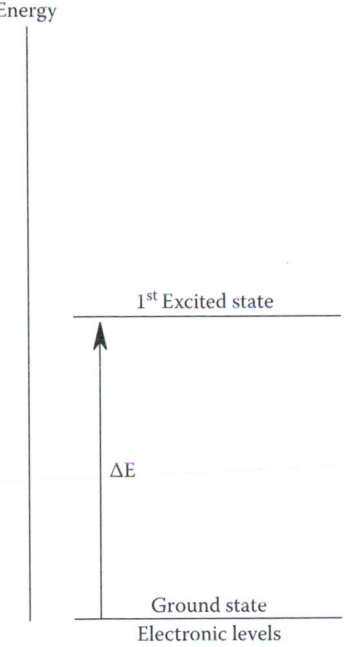

**Figure 5.3** Electronic transition from ground state to first excited state.

**Note:** The longer the wavelength is, the less energy the photon will have and vice versa.

Now let us consider what this means for molecules that we will be analysing: Each molecule of a substance has an internal energy that is the combination of the different energy associated with the molecules (the sum of the energy of its electrons, the energy of vibration between its constituent atoms). The energy levels of a simple molecule are shown in Figure 5.3. These energy levels are broadly separated and usually only the absorption of a high-energy photon can make the molecule move from one level to another. This becomes slightly different in complex molecules because the energy levels are more closely spaced and thus photons of visible light (and near UV light) can effect the transition (from one level to another) because less energy is required:

$$E = \frac{hc}{\lambda}$$

In comparison, the vibrational energy levels of the parts of the molecule are closer together than the electronic energy levels (see Figure 5.4). Because the vibrational energy levels are closer together (in comparison to the electronic

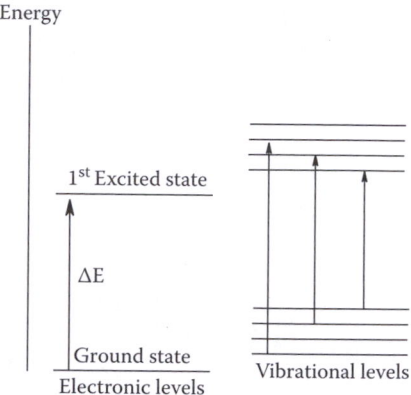

**Figure 5.4** Electronic and vibrational energy level transitions.

levels), photons of lower energy are sufficient to effect vibrational changes. Absorption of photons due only to vibrational energy changes is in the infrared region of the electromagnetic spectrum.

For UV and visible spectra, it would be expected from these theories that we would see a series of sharp lines. However, in practice, UV/Vis (ultraviolet/visible) spectra are rounded humps, as shown in Figure 5.5. This shows that the molecule is absorbing over a range of wavelengths rather than at one wavelength in particular. This range is due to the fact that the electronic level transitions are usually accompanied by a simultaneous transition between the numerous vibrational energy levels. It should also be noted at this point that each of the numerous vibrational energy levels associated with the electronic component has a smaller vibrational component and an even smaller rotational component.

When molecules are closely packed, such as when they are in solution, they will exert certain influences on each other that can affect the numerous energy levels and distort the sharp spectral lines into bands.

Now that we know how UV/Vis spectra are formed, we need to consider the types of substances that will absorb UV/Vis radiation and how we can identify them. It is true that, because we see coloured substances, they

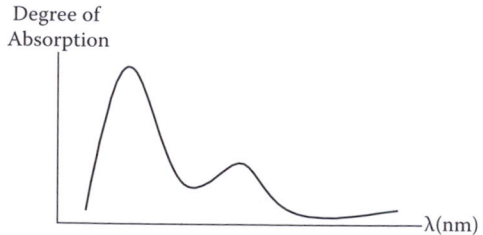

**Figure 5.5** Typical UV/Vis spectrum.

**Table 5.1    Wavelengths and Associated Colours**

| Emitted Wavelength (nm) | Colour Observed | Absorbed Wavelength (nm) | Colour Absorbed |
|---|---|---|---|
| 605–760 | Red | 490–500 | Blue-green |
| 595–605 | Orange | 480–490 | Green-blue |
| 580–595 | Yellow | 435–480 | Blue |
| 560–580 | Yellow-green | 400–435 | Indigo |
| 500–560 | Green | 380–400 | Violet |
| 490–500 | Blue-green | 605–760 | Red |
| 480–490 | Green-blue | 595–605 | Orange |
| 435–480 | Blue | 580–595 | Yellow |
| 400–435 | Indigo | 560–580 | Yellow-green |
| 380–400 | Violet | 500–560 | Green |

are absorbing or transmitting in the UV/Vis regions of the electromagnetic spectrum (see Table 5.1) and therefore we will be able to detect them using the visible detector. However, colourless substance can also be detected, so we need to understand what it is about these substances that makes them absorb these energies.

A relationship exists between the colour that a substance appears and its electronic structure, so the absorption of light energy (UV/Vis) will occur alongside an electronic state transition of the molecules in a sample.

## Bonding and Antibonding Orbitals

Consider the hydrogen ($H_2$) molecule shown in Figure 5.6. The hydrogen atoms, each having 1s-orbitals, will overlap to form the $H_2$ molecule. From this, we can see that the one molecular orbital is formed by adding the two 1s-orbitals together (it should be noted, however, that where the two orbitals overlap, their values will sum to give a larger result). In the $H_2$ molecular orbital, the electrons are found in the region between the nuclei where the electrons hold the nuclei together. Molecular orbitals that are concentrated

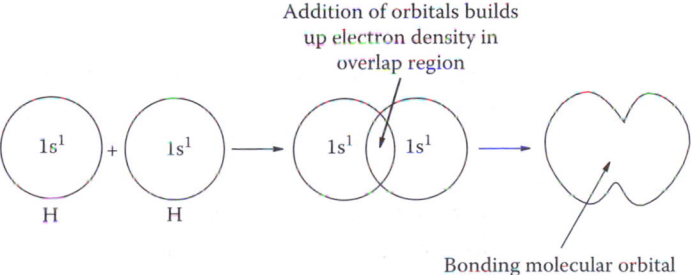

**Figure 5.6** Formation of $H_2$ bonding molecular orbital.

in the regions between the nuclei are called bonding orbitals, denoted by 'σ-bonding'. On the other hand, molecular orbitals that are concentrated in regions other than between two nuclei are called antibonding orbitals denoted by 'σ*-bonding'. Essentially, a number of different types of ground state orbitals may be involved in bonding:

σ (bonding) molecular orbitals as shown in C to C bonds:

π (bonding) as in a carbonyl C double bond to O:

η (nonbonding) atomic orbitals as in the presence of lone pair(s):

In addition, two types of antibonding orbitals may also be involved in the transition:

σ*-orbital
π*-orbital

**Note:** η*-orbitals do not exist because the electrons present in a η atomic orbital do not form bonds.

Antibonding molecular orbitals contain no electrons in the ground state of the molecule and are formed by the subtraction of orbitals, resulting in low electron density in the overlap region (Figure 5.7).

Consider the energy diagram for the combination of two hydrogen atoms to produce a $H_2$ molecule (Figure 5.8). Note that the two electrons occupying the bonding molecular orbital have lower energy than the sum of the two atomic orbitals that combined to form it. This is known as the *lowest electronic energy state* or *ground state* of the hydrogen molecule.

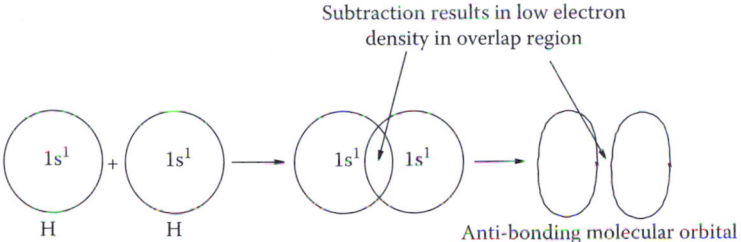

**Figure 5.7** Formation of $H_2$ nonbonding molecular orbital.

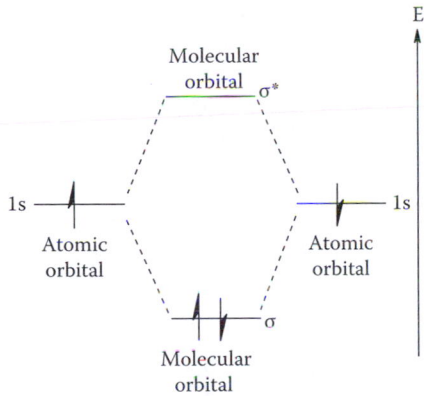

**Figure 5.8** Energy diagram for $H_2$ molecule.

**Note:** It is possible for an electron to occupy the antibonding molecular orbital; this is known as the excited state.

The strength of the interaction between the two atomic orbitals to form molecular orbitals is determined by the following two factors:

- the energy difference between the interacting orbitals
- the magnitude of the overlap between the orbitals

For a strong interaction to occur, the energy of both of the orbitals involved must be approximately equal and their overlap should be large. The example shown so far (hydrogen, $H_2$) has only shown the interaction of s-orbitals, but we also need to consider what happens in bonding when atoms have p-orbitals. There are two ways in which 2p-orbitals can interact to form bonds:

- One set of the orbitals can overlap along the axes to produce one bonding and one antibonding orbital ($\sigma_{2p}$ and $\sigma_{2p}^*$, respectively).
- The other set of p-orbitals can interact in a sideways fashion to produce two bonding and two antibonding π-orbitals (shown in Figure 5.9).

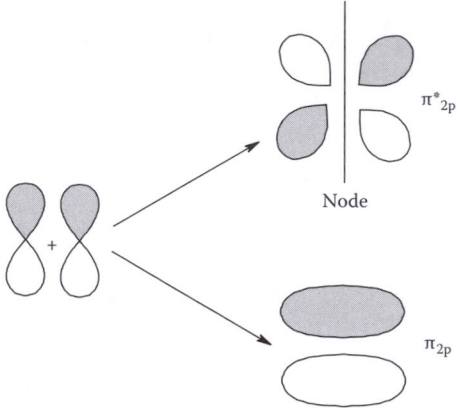

**Figure 5.9** **(See colour insert following page 142.)** p-Orbitals interacting to form two bonding and two antibonding π-orbitals.

The relative energies of the corresponding molecular orbitals formed from the interaction of two 2s- and two 2p-orbitals are shown in Figure 5.10.

Consider what this means for UV/Vis absorption detection methods: We know that by the absorption of energy (in this case, UV/Vis light), we find that transitions in which a bonding s-electron is excited to an antibonding σ-orbital are referred to as $\sigma \rightarrow \sigma^*$ transitions in the same way that an electron from a lone pair (nonbonding electron pair) to a $\pi^*$ orbital is called a $\pi \rightarrow \pi^*$ transition. The following transitions can occur following the absorption of UV/Vis light:

$\sigma \rightarrow \sigma^*$ (because this transition requires much energy, saturated groups
    do not absorb in the UV region)
$\eta \rightarrow \sigma^*$
$\eta \rightarrow \pi^*$
$\pi \rightarrow \pi^*$ (unsaturated molecules require less energy and transitions
    occur at longer wavelengths)

From this, it can be seen that the wavelength of maximum absorption ($\lambda_{max}$) is determined by the molecular structure of the molecule (due to the transitions occurring between the previously mentioned orbitals.

A $\pi \rightarrow \pi^*$ transition that occurs in an isolated group in a molecule, such as in the C=C bond ($\lambda_{max}$: 180–220 nm) will show absorption of low intensity. However, increased numbers of C=C bonds in a molecule (i.e., increasing unsaturated bonds, thus increasing conjugation within the molecule) leads to the $\lambda_{max}$ value increasing:

$\lambda_{max}$ = 180–200 nm          $\lambda_{max}$ = 258 nm          $\lambda_{max}$ ~ 280 nm

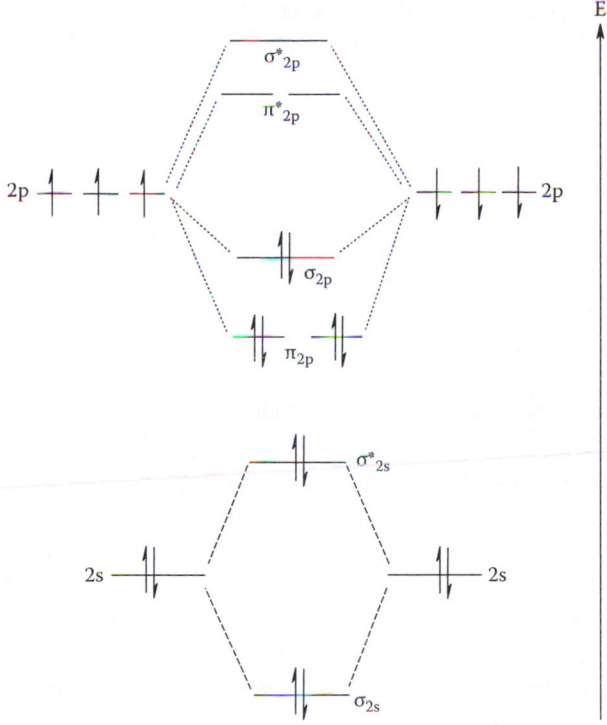

**Figure 5.10** Energy diagram for $N_2$ molecule.

In general terms, the more conjugated the molecule is, the closer to (and eventually within) the visible region of the electromagnetic spectrum the $\lambda_{max}$ will be.

Two types of groups can affect the absorption spectrum of a molecule: *chromophores* and *auxochromes*.

Chromophores are a group of atoms that are responsible for the absorption of UV/Vis light and principally give rise to the colour that a molecule exhibits (if any). For example,

— N = N —     azo

> = O     carbonyl

— N⁺ < O / O⁻     nitro

Commonly, electron withdrawing groups

Auxochromes are groups that are conjugated with the chromophore and 'enhance' or intensify the colour of a molecule. For example,

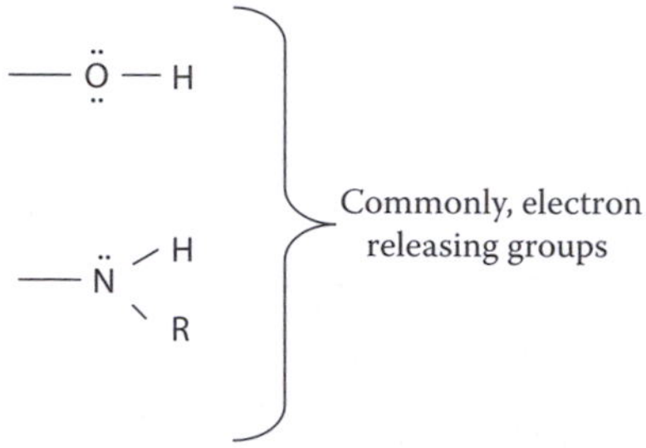

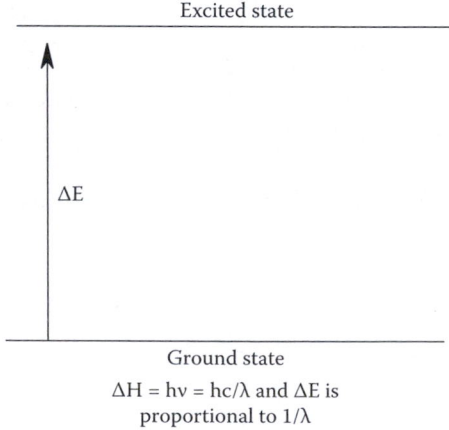

The colour of a molecule will arise from electronic transitions from a ground state to an excited state from the absorption of visible light. The wavelength of absorption is related to the change in energy ($\Delta E$) of this transition (see Figure 5.11).

The effect that an auxochrome will exert on a chromophore will depend on the polarity of the auxochrome. Some groups will have little effect on the wavelength, whereas others will produce distinct shifts in the wavelength absorptions (see Figure 5.12 and Table 5.2). It is also possible that bulky groups attached to the molecule will produce steric hindrance factors that may have to be considered.

Excited state

$\Delta E$

Ground state

$\Delta H = h\nu = hc/\lambda$ and $\Delta E$ is proportional to $1/\lambda$

**Figure 5.11**  Energy change between ground state and excited state.

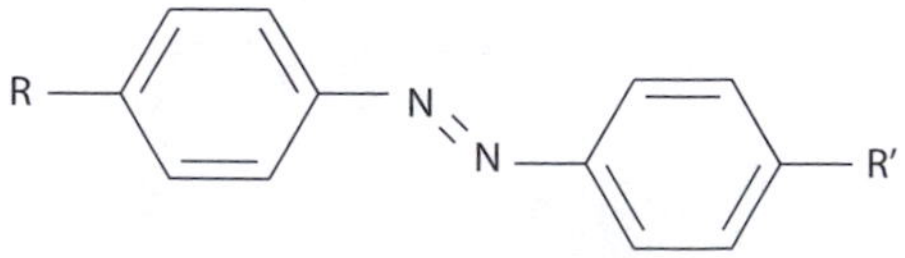

**Figure 5.12**  A20 compounds.

**Table 5.2 Wavelength Shifts Due to Auxochrome Differences**

| R | R' | $\lambda_{max}$ in Ethanol (nm) |
|---|---|---|
| H | H | 320 |
| H | $NO_2$ | 332 |
| $NO_2$ | $NO_2$ | 338 |
| H | $NH_2$ | 385 |

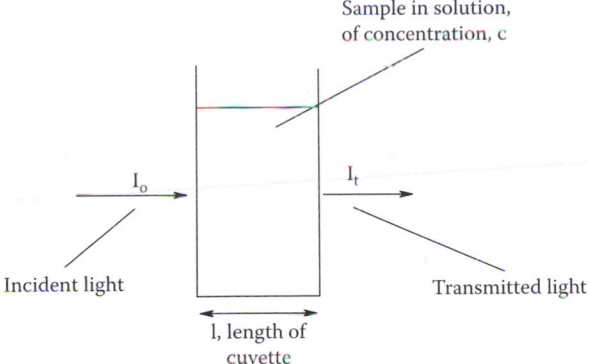

**Figure 5.13** The relationship between absorbed and transmitted light in UV/Vis spectroscopy.

In UV/Vis detection and quantitation of analytes, we are predominantly concerned with the absorption of the analyte; however, we must also consider that light is also being transmitted. Let us consider the absorption of the energy (UV/Vis light) by solutions containing our analytes of interest (Figure 5.13).

Equal amounts (fractions) of incident radiation are absorbed by equal changes in the cuvette cell length, with the concentration of the analyte remaining constant. If we consider what happens when we have a much smaller cuvette (Figure 5.14),

$$-dI \, \alpha \, I.dl$$

OR

$$-dI = k.I.dl \ (k = \text{proportionality constant})$$

Rearranging this gives

$$k.dL = \frac{-dI}{I}$$

and, by integrating,

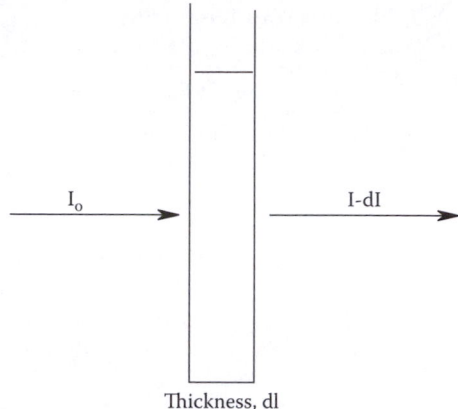

**Figure 5.14** The relationship between path length and absorbed light in UV/Vis spectroscopy.

$$-\int_{I_o}^{I_t} \frac{dI}{I} = k \int_{o}^{1} dl$$

Taking the natural log (ln),

$$[-\ln I]_{I_o}^{I_t} = k[l]_{o}^{l}$$

Then, by substituting the limits,

$$-\ln I_t - \left(-\ln I_o\right) = kl$$

Therefore, we arrive at Lambert's law:

$$\ln I_o - \ln I_t = kl$$

$$\log_{10}\left(\frac{I_o}{I_t}\right) = k'l$$

We know that equal amounts of incident radiation are absorbed by equal changes in the absorbing species (analytes) where the path length remains constant, so by parallel deviation to Lambert's law, we get Beer's law:

$$\log_{10}\left(\frac{I_o}{I_t}\right) = k''c$$

Combining the two laws gives the Beer–Lambert law:

$$\log_{10}\left(\frac{I_o}{I_t}\right) = k.c.l$$

It should be noted here that in most versions of the Beer–Lambert law, $k$ is replaced with $\varepsilon$, the molar absorption coefficient. $\log_{10}(I_o/I_t)$ is a function, known as absorbance ($A$), which results in the final equation:

$$\boxed{A = \varepsilon c l}$$

where

$A$ = absorbance

$\varepsilon$ = molar absorption coefficient ($dm^3\ mol^{-1}\ cm^{-1}$, which can also be written as $m^2\ mol^{-1}$)

$c$ = concentration ($mol\ dm^{-3}$)

$l$ = path length (cm)

Alternatively, if the relative molecular mass of the analyte is known, a 1% w/v solution can be prepared and the absorbance measured in a 1 cm cell. In this case, $\varepsilon$ is replaced by $E^{1\%}$, sometimes with the wavelength included, for example, $E^{1\%}$ (254 nm).

For quantitative analysis, a wavelength at which $\varepsilon$ is maximum ($\lambda_{max}$) will be chosen for the following reasons:

- The change in absorbance for a particular concentration change is greater, which will lead to a greater sensitivity and accuracy in the measurement.
- The relative effect of impurities or other substances will be smaller.
- The rate of change of absorbance with wavelength is smaller and the measurement will not be adversely affected by small errors in the wavelength setting.

A major factor in analyses, apart from the chemical nature of the analyte, is the solvent in which the analyte is dissolved. Because normal phase chromatography is rarely used nowadays, we will consider the solvents used in reverse phase chromatography. Table 5.3 shows solvents commonly used in reverse phase chromatography and their ranges of absorption.

In general, a solvent with a UV cut-off higher than the working wavelength used for analysis will generate such a high background absorbance that it should be excluded from further consideration because it will not be a practically suitable solvent.

**Table 5.3 Solvents Commonly Used in Reverse Phase Chromatography and Their Ranges of Absorption**

| Solvent | UV Absorption Cut-off (nm) |
|---------|:--------------------------:|
| Water | 180 |
| Acetonitrile | 190 |
| Tetrahydrofuran | 212 |
| Methanol | 205 |
| Isopropanol | 205 |

In UV/Vis detectors, a solvent that has a UV cut-off at a wavelength of 190 nm and below (i.e., does not absorb) would be the best solvent to use. Acetonitrile is a particularly good solvent to use for these reasons; however, at the time of writing, the worldwide shortage of acetonitrile has meant that many chromatographers have had to resort to alternatives. Sadly, this is not as easy as directly substituting another solvent for acetonitrile, such as methanol, because it has a cut-off of 205 nm, which means that it will produce noisy spectra (in this region) and may obstruct the absorbance spectra of analytes that absorb at these wavelengths. Other issues that may occur are change in backpressure of the column and 'salting out' (this can occur as precipitation in the pump if the concentration of a buffer, such as phosphate, is high and mixes with methanol).

**Note:** The remainder of this section outlines methods of reducing the amount of acetonitrile used and can be skipped if it is not necessary to the reader.

The amount of acetonitrile can be reduced by reducing the column length, internal diameter, and particle size. It is also possible, depending upon the analyte being detected, to change the solvent for another. However, this requires careful consideration and would require revalidation in a regulated environment, such as is found in forensic science and in the pharmaceutical industry (see Chapter 9).

If you choose to use an alternative column, the following equation can be used to determine the flow rates that would be required with the new column (note that this is true for all solvents and is not specific to acetonitrile):

$$F_2 = F_1 \times \left( \frac{L_2}{L_1} \right) \times \left( \frac{ID_2}{ID_1} \right)^2$$

where
$F_1$ = flow rate for original column
$F_2$ = flow rate for new column

$L_1$ = length of original column
$L_2$ = length of new column
$ID_1$ = internal diameter of original column
$ID_2$ = internal diameter of new column

Consider the following: You already have an established method using a mobile phase of methanol and water but you wish to reduce the amount of solvent that you are using, so you choose to change the column. The original column was a C18 (4.6 × 250 mm, 5 μm) with a flow rate of 1.5 mL min⁻¹. The new column you are considering is a C18 (2.1 × 150 mm, 5 μm). To determine the new flow rate,

$$F_2 = 1.5 \times \left( \frac{150}{250} \right) \times \left( \frac{2.1}{4.6} \right)^2 = 0.187$$

Therefore, a flow rate of 0.2 mL min⁻¹ (≡ 200 μL min⁻¹) would be used. If the run time of the original method is 20 minutes, this would produce solvent waste of approximately 30 mL. On the other hand, for the new column, the amount of solvent waste produced would be approximately 4 mL, which is a 7.5 times reduction in solvent waste. On an industrial scale, this is a huge saving of solvent.

It is also possible to keep the length of the column the same but to reduce the particle size:

$$SF = \frac{\left( ID_{small} \right)^2}{\left( ID_{large} \right)^2}$$

where $SF$ = scaling factor.

If we consider that the original method uses a C18 (4.6 × 250 mm, 5 μm) with a flow rate of 1 mL min⁻¹, and an alternative particle size is available (4.6 × 250 mm, 3.5 μm),

$$SF = \frac{3.5^2}{5.0^2} = 0.49$$

This equates to approximately 50% reduction in the flow rate (from 1.0 to 0.5 mL min⁻¹). Refer to Chapters 6 and 10 on method development and troubleshooting, respectively.

## UV/Vis Detectors versus Diode Array Detectors

In UV/Vis and diode array detectors, the analytes elute from the column in the flow of the mobile phase where it enters a flow cell inside the HPLC instrument. The flow cell is held in the path of a beam of UV/Vis radiation and the detector detects analytes that absorb in this region of the electromagnetic spectrum.

### UV/Vis Detector

This is the simplest form of UV/Vis detection; the source is produced by a deuterium, mercury, and/or tungsten filament lamp. A monochromator is used to produce a single narrow band of UV or visible radiation, which is passed through the sample and the transmitted radiation is detected by a photomultiplier. This response from the detector is linear with concentration, obeying the Beer–Lambert law as previously discussed. UV/Vis detectors can function in one of two ways:

- Fixed wavelength detectors are such that only wavelengths of radiation can be passed through the sample and detected. Typical wavelengths for these instruments are 220, 254, 436, and 546 nm and are achieved by using different source lamps (e.g., 254 nm, Hg lamp). This type of detector is used in specific applications where analyses require only one wavelength for detection.
- Variable wavelength detectors can provide one wavelength within the working range of the instrument.

### Diode Array Detector

Diode array detectors (DADs) provide more spectral information than UV/Vis detectors because the broad UV/Vis range of wavelengths can be passed through the sample. This produces a full UV/Vis spectrum for each analyte detected, as opposed to an absorbance value from the single-wavelength UV/Vis detectors. This type of detector is the most commonly used detector in forensic science labs because of the flexibility of being able to analyse substances eluting from the column over a wide range of wavelengths at the same time.

## Electrochemical Detectors

Electrochemical detectors are used in ion chromatography systems. It is not possible to observe UV/Vis absorption in the typical ions analysed in ion chromatography; therefore, alternative detection methods must be used. Two types of detectors can be found in electrochemical detection: conductance detectors and amperometric detectors.

## Conductivity Detectors

This type of detector is used in ion chromatography for the detection of inorganic anions (e.g., $SO_4^{2-}$, $PO_4^{3-}$), some inorganic cations (e.g., $Ca^{2+}$, $Mg^{2+}$), and some ionised organic acids. This is due to the fact that all ions are electrically conducting. Conductivity detectors are based on the conductance of an eluent prior to and during the elution of the analyte from the column.

## Amperometric Detectors

This type of detector measures the current generated in an electrochemical cell at a fixed applied potential by the reduction or oxidation and an eluted analyte at the surface of a microelectrode. Sometimes this is called the counterelectrode (usually made of gold, platinum, or a glassy carbon) and auxiliary or working electrode and a reference electrode, usually Ag/AgCl. The mobile phase used will act as the supporting electrolyte for the redox reactions; therefore, its composition is restricted to aqueous solvent mixtures. Amperometric detection is used for ions that have pKa values of greater than 7 and thus cannot be detected by conductivity detectors (because the formed products are weakly dissociated).

The potential required or the reduction or oxidation of the analyte being detected is applied between the auxiliary and the reference electrode. The microelectrode acts as the counterelectrode and functions to protect the potential and to prevent damage to the reference electrode. When a mixture containing ions (electrochemically active) flows through the measuring cell within the instrument, it is partially reduced or oxidised. This in turn produces a cathodic or anodic current that is proportional to the concentration of the analyte.

## Mass Spectrometry

Mass spectrometry (MS) is an established technique in its own right and has found many applications, especially in chemical analysis and in research. Used on its own, it is a very powerful technique that provides mass data for a compound. When used in conjunction with other instrumentation, this can help to elucidate the structures of newly synthesised compounds or can provide quality data for compounds in the pharmaceutical industry. It is used for single component, pure samples.

Mass spectrometry, on its own, does not find much, if any, use in forensic science due to the nature of the samples that are analysed (sample matrices tend to very complex, e.g., blood). In this section, we shall look at the aspects of mass spectrometry as it relates to hyphenation with liquid chromatography. Hyphenation is the combining of the two techniques whereby HPLC is used to separate a mixture of components; each analyte is introduced into the mass spectrometer as a single compound (i.e., no longer part of the matrix of the original sample).

The first step of the analysis of a sample is its introduction to the MS instrument: A pure sample can be introduced directly by dissolving the substance in a small amount of an appropriate solvent and injecting it. It is also possible to introduce solid samples through an inlet. This is not the case when we have a hyphenated technique because usually we have the issue of relatively high flow rates of organic solvent (from the HPLC instrument) and these cannot be introduced to the MS instrument. If it is necessary for the LC method to stay the same (with high flow rates), an interface will have to be employed to reduce the chromatographic mobile phase.

As we have already seen in our previous discussions of UV/Vis detection, columns can be changed, thus reducing the flow rate of the mobile phase. For this reason, small bore columns (<4.6 mm internal diameter) are generally used when using LC-MS to avoid the extra step of mobile phase solvent reduction.

As with all of the detectors we have encountered so far, variation is possible in the separate component of the MS instrument (as shown in Figure 5.15). The variations that we will look at will be those specific to the instruments that can be hyphenated with liquid chromatograph instruments. For further information relating to alternatives, please see the 'Further Reading' section at the end of this chapter.

Figure 5.15 outlines the basic steps in mass spectrometric analysis; because we have already mentioned the sample introduction step, we shall consider the other component.

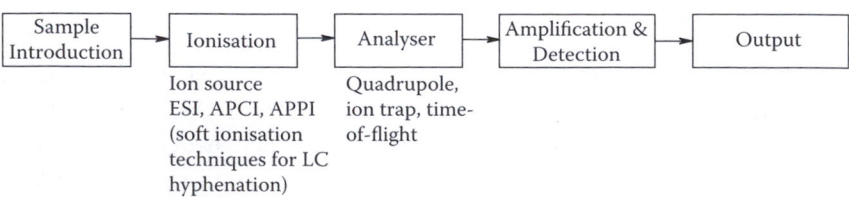

**Figure 5.15** Block diagram of a mass spectrometer.

## Ion Source

The ion source produces ionisation of the analyte as it enters the mass spectrometer. Ionisation can be brought about in a number of ways. However, the so-called 'soft' ionisation techniques used in LC-MS systems are electrospray ionisation (ESI), atmospheric pressure chemical ionisation (APCI), and, more recently, atmospheric pressure photoionisation (APPI).

### *Electrospray Ionisation (ESI)*

Electrospray ionisation generates analyte ions; this is accomplished by spraying the eluent (mobile phase solvent + any analytes eluting from the LC system) into a chamber at atmospheric pressure. This is done in the source in the presence of a heated drying gas (usually $N_2$) and a strong electrostatic field. The pressure of the electrostatic field causes further dissociation of the analyte molecules and the drying gas causes the solvent to evaporate (see Figure 5.16).

On evaporation of the solvent, the charge concentration of the droplets increases, which in turn leads to the forces between ions of the same charge ejecting ions into the gas phase (known as desorbing). These ions then pass through a capillary chamber, as shown in Figure 5.17, and into

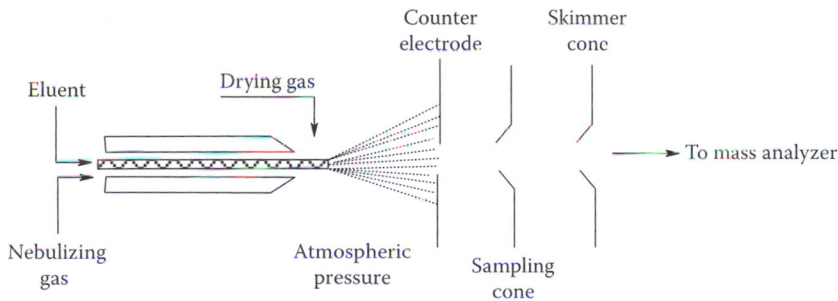

**Figure 5.16** ESI ion source.

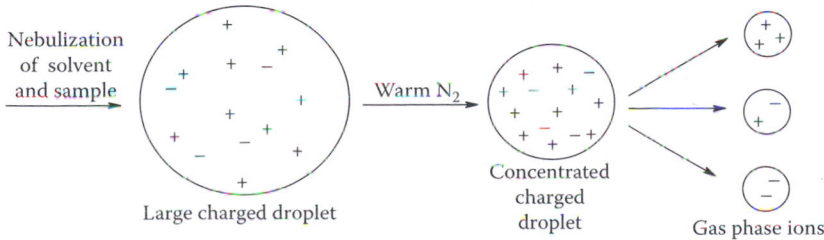

**Figure 5.17** Ion formation within ESI source.

the mass analyser. Electrospray ionisation is useful for the analysis of large biomolecules due to its wide mass range (60 Da–150 kDa) and is also used in the analysis of drugs, toxicological samples, explosives, and environmental samples.

### Atmospheric Pressure Chemical Ionisation (APCI)

The main difference between ESI and APCI is that, in APCI, the ions are formed when the analyte molecules in the gas phase interact with electrons discharged from a corona needle (see Figure 5.18). As with ESI, the eluent is sprayed through a heated vaporiser at atmospheric pressure. The heat from the vaporiser causes the liquid to vaporise and the resultant solvent molecules, now in the gas phase, to be ionised by the electrons from the corona needle. These ions then transfer charge to the analyte molecules through chemical reaction (hence the name 'atmospheric chemical ionisation'). These ions, as in ESI, pass through a capillary and pass into the mass analyser.

### Atmospheric Pressure Photoionisation (APPI)

This is a relatively recent addition to the ionisation techniques used with hyphenated liquid chromatography. In this type of detector, a vaporiser converts the eluent (from the LC) into the gas phase, much like with APCI. The difference with this technique is that instead of the production of electrons from a corona, here we have a discharge lamp producing photons (known as vacuum ultraviolet photons) in a narrow range of ionisation energies.

### Analyser

The analyser is the part of the mass spectrometer that filters the fragments and ions produced in the ion source before they are detected and amplified and the output viewed. Again, a number of different types of analysers are

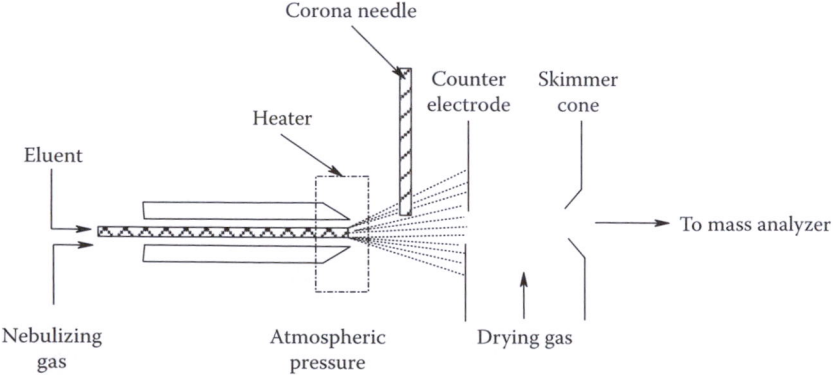

**Figure 5.18** Ion source of APCI.

available on the market for mass spectrometers; however, we shall cover only those that can be applied to hyphenated LC systems. These include the quadrupole, the ion trap, and the time-of-flight (TOF) analysers. We shall also cover tandem mass spectrometry because this technique is finding its place in forensic science laboratories.

### Quadrupole

The quadrupole is an analyser that detects over a wide range of masses (range), provides good reproducibility of mass spectra, and can be used for qualitative and quantitative analyses. The quadrupole is a series of four parallel rods that are equally spaced around a central area, as represented in Figure 5.19.

The rods of the instrument have an electrical potential applied to them, with opposing rods having the same charge and adjacent rods having different charges (represented in Figure 5.19). The fragments and ions travel through the centre of the rods and only ions with a certain mass-to-charge ratio (m/z) will be allowed to pass through into the detector. The rest of the fragments and ions will collide with the walls of the rods and will be dissipated. Voltages can be applied to the rods in order to detect a certain range of m/z values (full scan monitoring) or to monitor specific m/z values of interest (single ion monitoring [SIM]). This type of analyser can be used with m/z values up to 4,000 Da and therefore can be used with ESI, APCI, and APPI ion sources.

### Ion Trap

This type of analyser consists of a doughnut-shaped electrode and two other hyperbolic electrodes, which are essentially end-caps (shown in Figure 5.20). The ions are trapped in the centre of the ring electrode by applying RF potentials whereby an electrostatic ion gate pulses open and closed to inject the ions, and by having the ion trap filled with a damping gas. This is typically helium (~1 mtorr or ~0.1 Pa), which is used to collide with the ions, thus reducing their kinetic energies and allowing contraction of trajectories (of the ions) toward the centre of the analyser. An ion will be trapped and ejected to the detector, depending on its mass-to-charge ratio; this is done by selec-

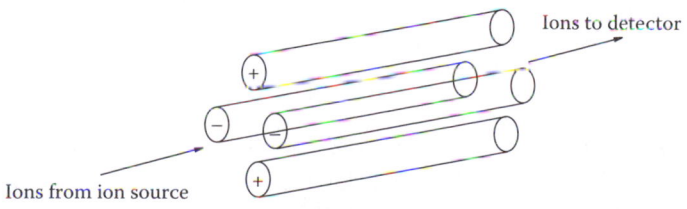

**Figure 5.19** Quadrupole analyser.

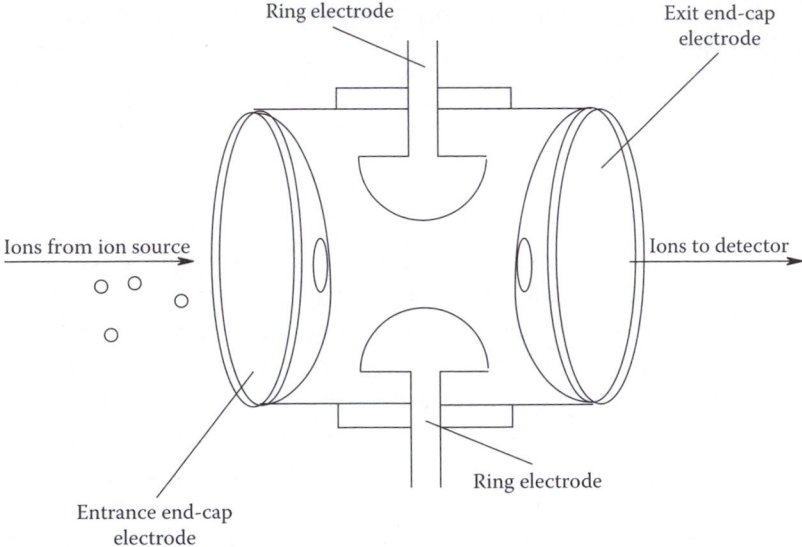

**Figure 5.20** Ion trap analyser.

tively altering the RF voltage of the trap. When a scan cycle is complete, other ions will be introduced to the trap and the process repeated.

There are some limitations of the ion trap analyser in that it is subject to *space-charge effects*. These effects are a consequence of too many ions present in the trap, which results in a distortion of the electrical field. This in turn leads to a reduction in the overall performance of the analyser. This type of analyser can be used up to m/z values of up to 4,000 Da, as with the quadrupole. The main difference between the quadrupole and the ion trap analyser is that quadrupoles act as a continual mass filter with ions continuously transferred to the quadrupole from the ion source, whereas ion trap analysers selectively trap ions of certain m/z values.

### Linear Ion Trap (Quadrupole Ion Trap)

The linear ion trap analyser is a variation of the ion trap analyser introduced to combat some of the problems that had been encountered with the ion trap in relation to the space-charge effects. The linear ion trap utilises two trapping electrode rings on the end of a quadrupole. This allows the instrument to be used in a normal scanning quadrupole mode; it also allows for the end electrodes to be switched on to retain only specific ions in the trap for collision with the damping gas.

The major difference between these analysers and normal ion trap analysers is that they have a much greater capacity for trapping within the volume between the rods of the quadrupole (as opposed to in the centre of the ring electrode or the normal ion trap). This results in a greater sensitivity for analysing effluent from the HPLC for minor components, such as metabolites in toxicological analyses.

### Time of Flight (TOF)

This type of analyser measures the time that it takes for an ion or fragment to reach the detector, which is at a known distance from the source. The time detected is dependent on the m/z of the ion. Thus, by detecting the time taken for the ion to travel the distance to the detector, the m/z can be determined. This type of analyser has been used with ESI sources.

### Tandem Mass Spectrometry

Tandem mass spectrometry is a combination of mass spectrometry analysers. An example of this is a triple quadrupole instrument (MS/MS) where three quadrupoles are in series. The first quadrupole acts as an analyser to detect the ions from the ion source, the second acts as a collision cell, and the third acts as an analyser to detect the fragments that are formed in the collision cell. Inside the collision cell, an inert gas (usually argon or xenon) is introduced; the gas will collide with the ions from the sample and cause them to further fragment.

## Amplification and Detection

After the ions have been separated by the analyser, they will be focused onto the detector, where they are converted into a measurable electrical current. This results in a signal in the form of a series of peaks showing the abundance of those particular ions. The most common type of detector is an electron multiplier, which exists in two forms: the discrete dynode electron multiplier and the continuous dynode electron multiplier (also called a channel electron multiplier [CEM]).

In the discrete dynode electron multiplier, the ions from the analyser are converted into electrons by a dynode (an electron used to provide secondary emission). The dynode surface is typically composed of CsSb, BeO, or GaP, which are secondary emitting materials. This means that the electrons are emitted or released from atoms in the surface layer with the number of electrons released depending upon

the type of primary particle (ions)
the energy of the primary particle
the characteristic of the incident surface (or the dynode)

The amplification of the signal is achieved by having a series of dynodes, usually 12 or 24, that have increasing potentials applied and maintained; the

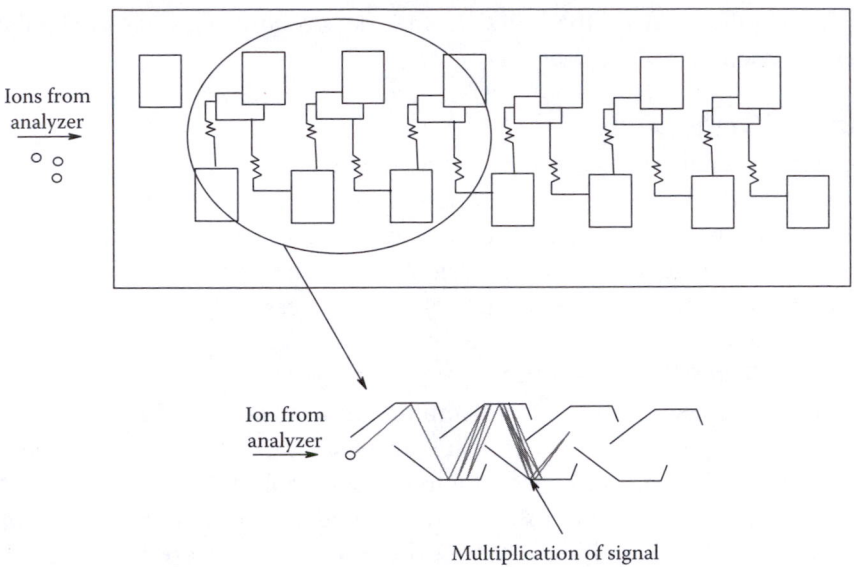

**Figure 5.21** Discrete dynode electron multiplier.

secondary emissions of electrons produced in each of the dynodes is further multiplied in the next (see Figure 5.21).

In comparison, the continuous dynode electron multiplier differs in that the amplification of the signal occurs by the electrons colliding with the internal surface of the detector. The detector is a continuous dynode that is horn shaped; it is shown in Figure 5.22.

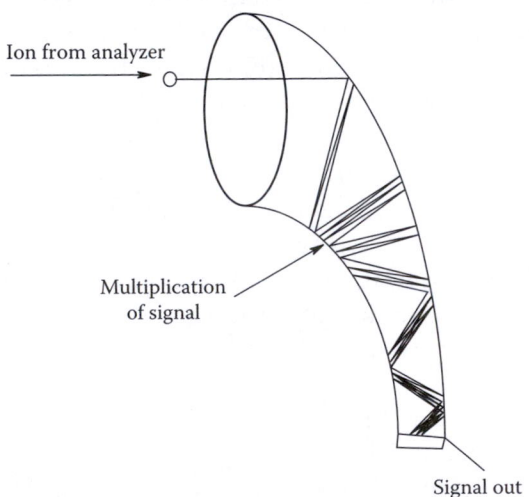

**Figure 5.22** Continuous dynode electron multiplier.

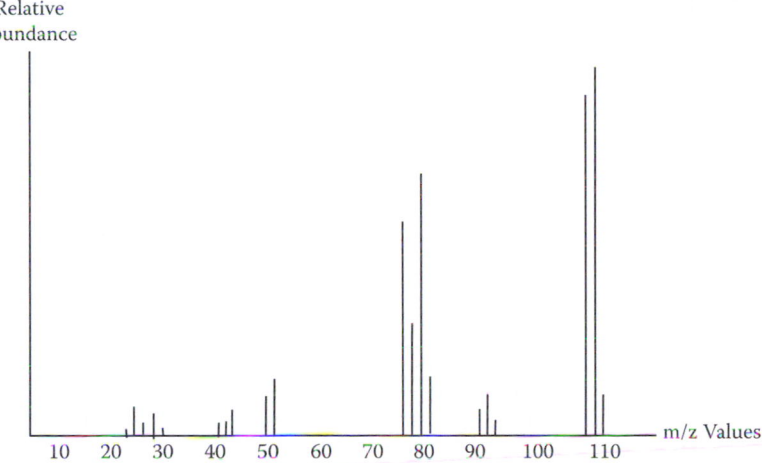

**Figure 5.23** Example of a mass spectrum.

## Output

The resulting spectra will be a series of peaks representing the abundance versus m/z value, as shown in Figure 5.23.

Depending upon the different instrument used and the manufacturer from which the instrument was purchased, it is possible to manipulate the spectra obtained to a certain degree. It is possible to obtain information on the relative abundances of each of the m/z values in the scan range or selected ions.

In forensic science, the most common type of system used is an LC–ESI–MS, usually with a single or triple quadrupole; however, ion trap is also finding its way into use. Most of these instruments will also have DAD detection, which can be switched on or off as necessary.

---

**KEY POINT SUMMARY**

$$E = \frac{hc}{\lambda}$$

Beer–Lambert law: $A = \varepsilon c l$

A *chromophore* is a group of atoms responsible for the absorption of UV/Vis light and principally gives rise to the colour that a molecule exhibits (if any). Examples include:

*(continued on next page)*

$$-\text{N} = \text{N} - \quad \text{azo}$$

$$\text{C} = \text{O} \quad \text{carbonyl}$$

$$-\text{N}^+ \overset{\text{O}}{\underset{\text{O}^-}{\diagdown}} \quad \text{nitro}$$

} Commonly, electron withdrawing groups

*Auxochromes* are groups that are conjugated with the chromophore and 'enhance' or intensify the colour of a molecule. Examples include:

$$-\ddot{\text{O}} - \text{H}$$

$$-\ddot{\text{N}} \overset{\text{H}}{\underset{\text{R}}{\diagup}}$$

} Commonly, electron releasing groups

Transitions that can occur following the absorption of UV/Vis light include:

$\sigma \rightarrow \sigma^*$

$\eta \rightarrow \sigma^*$

$\eta \rightarrow \pi^*$

$\pi \rightarrow \pi^*$

UV/Vis detection: utilising the absorption (or transmittance) of UV/Vis light by an analyte that is represented in a spectrum. The response from the detector is proportional to the concentration of the analyte (however, there are exceptions to this rule).

Electrochemical detection: these types of detectors are based on the conductance of an eluent prior to and during the elution of the analyte from the column (conductance) or the current generated in an electrochemical cell at a fixed applied potential by the reduction or oxidation and an eluted analyte (amperometric).

Mass spectrometry: the analyte is broken up into fragments or ions resulting in a spectrum of m/z ratio versus percent abundance.

## QUESTIONS

1. Which of the following would absorb in the UV/Vis region of the electro-magnetic spectrum?

Hydrocortisone

CI. Disperse Orange 25

Gabapentin

2. Which of the following 2 chemical compounds would absorb at the longest wavelength?
3. Identify the chromophore/auxochrome in each of the following:

4. Given the following data, determine the molar absorption coefficient ($\varepsilon$) of the analyte:

| Concentration (mg/mL) | Absorbance |
|---|---|
| 10 | 0.101 |
| 20 | 0.200 |
| 30 | 0.302 |
| 40 | 0.401 |
| 50 | 0.500 |
| Unknown sample (run 1) | 0.366 |
| Unknown sample (run 2) | 0.370 |
| Unknown sample (average) | 0.368 |

*Note:* Path length of cuvette = 1 cm.

5. Given the following scenarios, which detector would you choose for the analysis?
   a.  Determination of the concentration of calcium in bottled water
   b.  Detection and quantitation of diazepam and desmethyldiazepam
   c.  General screen of a urine sample for the presence of drugs of abuse
   d.  Determining the concentration of dye added to a fizzy drink

# Further Reading

Christie, R. M. 2001. *Colour chemistry.* Cambridge, England: Royal Society of Chemistry.
Christie, R. M., R. R. Mather, and R. H. Wardman. 1999. *The chemistry of colour application.* Chichester, England: Wiley-Blackwell.
De Hoffmann, E., J. Charette, and V. Stroobant. 2007. *Mass spectrometry: Principles and applications,* 3rd ed. Chichester, England: Wiley-Blackwell.
Flanagan, R. J., D. Perrett, and R. Whelpton. 2005. *Electrochemical detection in HPLC: Analysis of drugs and poisons.* Cambridge, England: Royal Society of Chemistry.

# Method Development in Reversed Phase HPLC

# 6

## Introduction

Method development in HPLC is a challenging and interesting field of study. A number of factors need to be addressed in any method development exercise. We will deal with them on an individual basis in this chapter.

The majority of analyses performed by HPLC in forensic science can be covered by using RP (reversed phase)-HPLC in one form or another. Column chemistry has advanced significantly, thus expanding the scope of such a technique. As a result of this, we have limited most of our method development discussions to reflect this. Additional texts covering method development for alternative separation modes can be found in the 'Further Reading' section at the end of this chapter.

When a method for HPLC is developed, it should be noted that each of the parameters that will be discussed is in no way isolated from one another and that any changes in the process should be made in a systematic manner. Automated systems are available to assist with the method development process and examples of these will be discussed as well. However, before any software programme is used, it is necessary to have an understanding of the processes involved in order to ensure that a correct and valid outcome is achieved.

The three most important things that need to be addressed are

- the nature of the analytes to be analysed (sample composition)
- the type of column that will effect the separation (choosing a column)
- the mobile phase that will achieve the best separation (choosing a mobile phase)

## Sample Composition

We already know that the chromatographic separation process depends on the interaction or affinity of the analytes of interest with the stationary and mobile phases (see Chapter 2). Therefore, it is important that we know or find out the answers to the following questions about our analytes of interest.

## What Is the Sample Matrix? Will It Interfere with Our Chromatography?

Chapter 3 dealt with sample preparation in more detail. In forensic science, we need to consider the likelihood of interference due to the sample matrix.

### EXAMPLE 6.1

Toxicology samples that we might need to consider may include blood or urine, both of which contain the analytes of interest in a complex matrix that might interfere in an HPLC analysis.

Some of the components of the matrix may elute at the same retention time as the analyte, resulting in an incorrect measurement or estimation in relation to the analyte of interest in terms of both qualitative and quantitative measurement. It is important, therefore, that we acknowledge the sample matrix during our method development process so that we can try to ensure adequate exclusion or separation between it and our analytes of interest.

This is important so that we are able to determine whether the analyte is polar or nonpolar in nature and to recognise the shape of the molecule and the functional groups that are associated with the polarity (see Table 6.1). The structure will also provide information about whether the analyte is acidic or basic.

## What Is the Molecular Weight of the Analyte of Interest?

The molecular weight is important when looking at larger molecules in particular because they do not chromatograph as well on conventional columns and require a packing material with a larger pore size. Analytes with similar chemistries will elute depending on their molecular weight (see the 'Method Development Worked Example' section concerning parabens) and the larger molecules will elute last. Most of the analytes found in forensic examinations fall into what is known as the small molecule category and would not require special treatment. Proteins and peptides are examples of the types of molecules that would require special columns.

## What Is the pKa of the Analyte of Interest?

The pKa or acid dissociation constant is a measure of the strength of an acid in solution. A larger value for the pKa tells us that the analyte dissociates to a lesser extent and is therefore a weak acid. For basic analytes, the reverse applies, in that a higher value for the pKa means a stronger base.

Knowing the pKa of the analyte helps us to identify which form (ionised or non-ionised) will be present at a particular pH. Using aspirin (an acid

**Table 6.1  Functional Groups and Polarity**

| Functional Group | Polarity |
|---|---|
| | Polar |

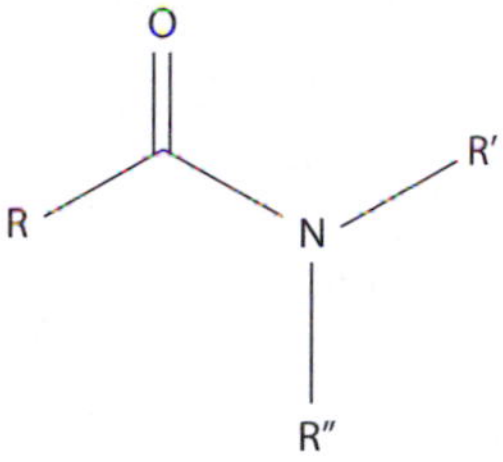

Amide

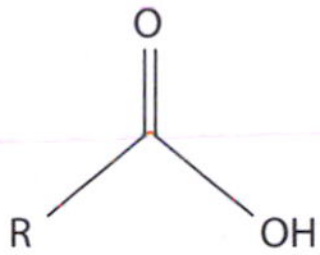

Carboxylic acid

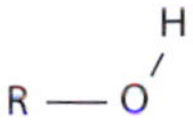

Alcohol

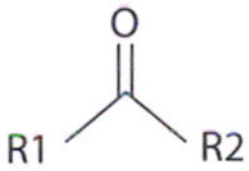

Ketone

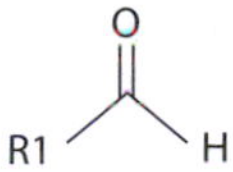

Aldehyde

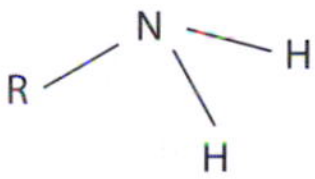

Primary amine

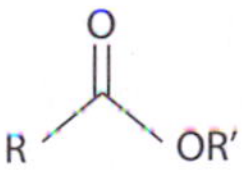

Ester

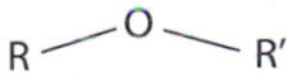

Ether

$R\text{-}C_n$                    Nonpolar

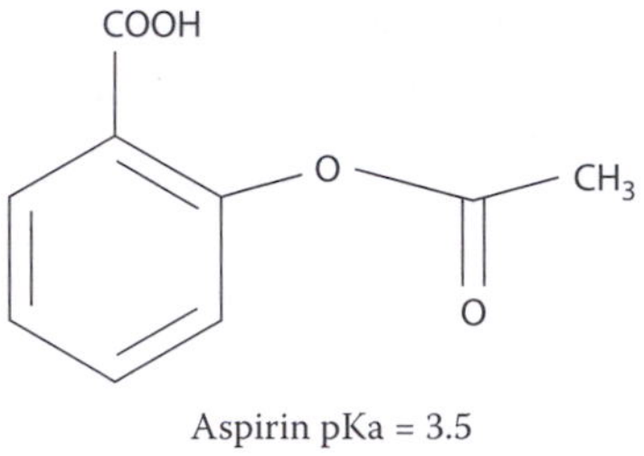

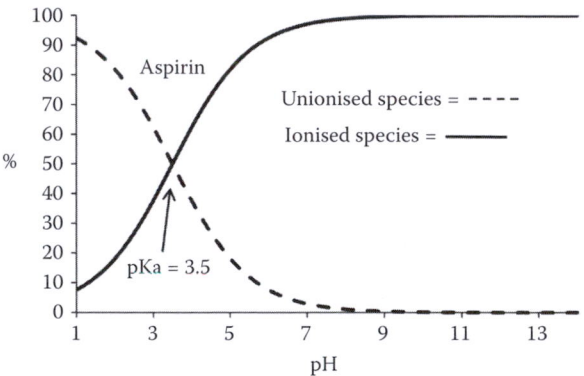

Aspirin pKa = 3.5

**Figure 6.1** Structure of aspirin.

**Figure 6.2** Ionisation of aspirin with change in pH.

with a pKa of 3.5) as an example (see Figure 6.1), we can deduce that at pH 2 the aspirin will be approaching the non-ionised form and at pH 6 it will be ionised (see Figure 6.2). In terms of chromatography, this process allows for some flexibility in relation to making our analytes of interest more or less polar, thus affecting retention. We need to ensure that we have adequate control over the mobile phase pH in order to ensure reproducibility in terms of retention time and good peak shape.

In RP-HPLC, if a small polar molecule is in the ionised form, you may find that it elutes at $t_M$ along with any other small polar molecule that may be present in the sample matrix. This would mean that the chromatography has been unsuccessful because no retention has taken place. In order to correct this, the sample would need to remain in the non-ionised form. This can cause problems when the analyte is very basic or very acidic because the pH required to ensure that the analyte remains in the non-ionised state is so extreme that the column cannot withstand the conditions and deteriorates rapidly. Operating with conventional columns at such extremes of pH both at the acidic end and alkaline end of the pH scale causes damage to the column.

## What Are the Detection Characteristics of the Analyte of Interest? For Example, Does the Analyte Absorb UV Light or Will Derivatisation Be Required?

Again, this relates back to the structure of the analyte. Understanding the nature of the functional groups present will help to give us an indication as to what detector would be suitable. This is discussed in Chapter 4, where we look at detection in greater detail.

## In Which Solvents Is the Analyte of Interest Likely to Be Soluble (Linked to Polarity)?

We know that the choice of mobile phase affects the separation and we will see later in this chapter why this is. Therefore, one of the things that we need to know is which solvents to use to dissolve the sample and to prepare the mobile phase. We can use the 'like dissolves like' rule of thumb when selecting a sample solvent.

### EXAMPLE 6.2

Polar analytes will be soluble in polar solvents and vice versa. If we think of oil and water, we know that the two substances do not mix because water is polar (high dipole moment) and oil is nonpolar. Let us look at a small alcohol; ethanol is soluble in water because the two analytes are polar in nature (see Chapter 2). The polar O–H bond in the ethanol structure dominates the relatively small, nonpolar C–H bonds, resulting in an overall polar molecule.

On the other hand, although octanol still retains the polar O–H bond, it now has a much larger nonpolar hydrocarbon grouping, which dominates the molecule, resulting in a nonpolar molecule that is insoluble in water (see Figure 6.3).

By ensuring that the analytes are soluble in the sample solvent, we can make sure that we obtain complete dissolution of the material, leading to a more accurate determination of the content. If the material is only partly soluble, then the measurement will be made only on the soluble portion of the material. This means that the results will be misleading and incorrect in terms of absolute concentration. Analytes that are either very nonpolar or very polar in nature are difficult to chromatograph by conventional HPLC.

A                                              B
$CH_3$ —— $CH_2$ —— OH                            OH
Structure of ethanol (**A**) and octance (**B**)

**Figure 6.3** Chemical structure of ethanol and octanol.

### EXAMPLE 6.3

Very nonpolar analytes are not soluble in solvents that are compatible with the types of aqueous and organic mixtures that might be expected to be used as mobile phases, and as such, tend to be analysed using gas chromatography.

Incompatible sample solvents and mobile phase may lead to the sample coming out of solution within the HPLC system.

### EXAMPLE 6.4

If we dissolve a highly water-soluble material in pure water and then introduce this to an HPLC system using 90% methanol as the mobile phase, we will find that the analyte of interest will precipitate out of solution because it is no longer soluble in the solvent mixture. At best, this causes a blockage; at worst, it leads to an inaccurate result for the same reasons as those mentioned earlier, which is not desirable. Therefore, solubility of the sample is an important parameter in deciding which mode of separation to use.

## Which Column Will Affect the Separation?

Here we need to consider which column might be the best one with which to start depending on our analytes of interest. For example, is the column going to need to be nonpolar in nature and, if so, would a $C_8$ suffice or is a $C_{18}$ required? The following parameters should be carefully considered.

### Separation Mechanism—RP-HPLC, NP-HPLC, Ion Exchange, Chiral, etc.

This choice will depend on the nature of the analytes that are being separated and the mechanism required to maximise the separation (see Chapter 4). For example, if the analytes are particularly acidic in nature, it may be necessary to choose ion exchange chromatography. If the analytes are nonpolar in nature, then reversed phase HPLC might be a suitable choice.

### EXAMPLE 6.5

Amphetamine (see Figure 6.4) is a weak base with a pKa of 10.1. It is basic in nature due to the $-NH_2$ group and is nonpolar due to the predominance of nonpolar C–C and C–H bonds. If we were to leave the pH and mobile phase composition aside for the moment, we would expect amphetamine to have an affinity for a nonpolar stationary phase and we would expect this affinity to increase with column packing chain length. Therefore, we can say that we would expect to achieve satisfactory chromatography using RP-HPLC (refer to Chapter 4).

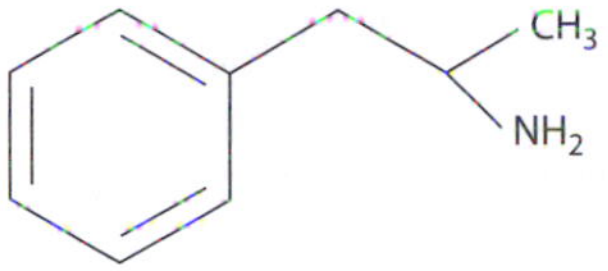

Amphetamine pHa = 10.1

**Figure 6.4** Structure of amphetamine.

## Column Selectivity—ODS, CN, NH$_2$, Silica, Anion/Cation, etc.

Column selectivity is based on the affinity of the analytes of interest and the stationary phase. If we take reversed phase chromatography as an example, we know from Chapter 4 that the interaction is based on the hydrophobicity of both the stationary phase and the analytes of interest. The more hydrophobic the analyte is, the more it will be retained on a hydrophobic column. Similarly, the more hydrophobic the column is, the more it will retain hydrophobic analytes.

Reversed phase stationary phases are generally composed of a silica backbone and a covalently bonded carbon/hydrogen chain ($-CH_2-CH_2-CH_2-CH_3$). An increase in the chain length is associated with an increase in the hydrophobicity of the packing material. Most column manufacturers supply columns with a range of chain lengths; the most frequently found include $C_4$, $C_8$, and $C_{18}$. As the chain length is increased, it would be expected that the retention time for a drug like amphetamine would increase. Hydrophobic analytes tend to be nonpolar in nature and we have already seen this in relation to amphetamine. If the column becomes more hydrophobic as the chain length increases, then the amphetamine will be more strongly retained by the packing material. This is due to the hydrophobic interactions and weak Van Der Waals forces, as discussed in Chapter 2.

We also know that analytes ionise at pH values above or below the pKa, depending on the nature of the species of interest (i.e., acidic or basic). Our choice of column, therefore, is dependent on the ionised state of the analytes at the time of analysis. Amphetamine has a weak base with a pKa of 10.1. In order to retain the analyte in the non-ionised form, the pH of the mobile phase would need to be approximately 11 or 12 (see Figure 6.5). This would provide an ideal condition for separation because there would be maximum retention of the hydrophobic amphetamine. However, in reality, traditional silica-based columns are not stable at such extremes of pH, so it is not practical to operate above pH 11. The silanol linkages are cleaved below pH 2 and the silica may dissolve at pH levels greater than pH 8.

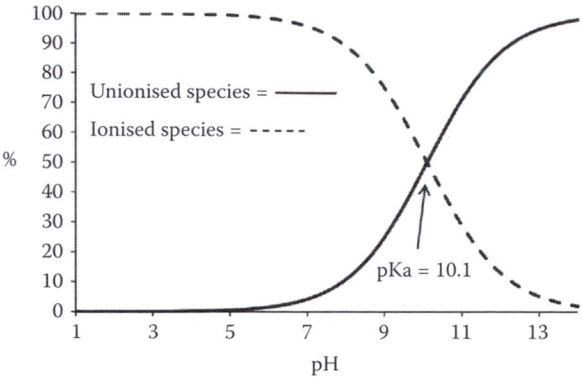

**Figure 6.5** Ionisation of amphetamine with change in pH.

**Table 6.2   Column Chemistry**

| Column Chemistry | Column Packing Type | pH Range | Application |
|---|---|---|---|
| See Figure 6.6 | End capped | 2–8 | Basic compounds |
| Bulky silane group with polar embedded | Polar embedded (bonus RP) | 2–8 | Basic analytes—reduced interactions with residual silanol groups |
| See Figure 6.6 | Bulky silane groups (stable bond) | 1–6 | Basic analytes |
| See Figure 6.7 | Polymeric | 2–13 | Strongly basic compounds |
| High degree of saturation of alkyl chain | Densely bonded phase (eclipse) | 2–9 | Acids, neutral, and bases |

Over the years, column technology has advanced and packing materials are now available that allow reversed phase separations at much wider extremes of pH. These columns have been chemically modified by a variety of means, including increasing the density of bonded phase, double end capping, and embedded polar groups within the stationary phase matrix (see Table 6.2 and Figures 6.6 and 6.7). These columns now offer greater flexibility in terms of the operational pH range.

These technologies provide greater stability and reduced dissolution of the packing materials at higher pH values. This means that basic analytes can be analysed using mobile phase pH levels above the pKa value, which improves retention and reduces peak tailing.

Referring to column manufacturers' literature is a good way of keeping up to date with column technologies; however, you should be aware that subtle changes in the column chemistry are often hailed as a breakthrough in technology. Even if your analytes of interest are relatively unknown (as can happen in forensic science), you can be sure that one of the column

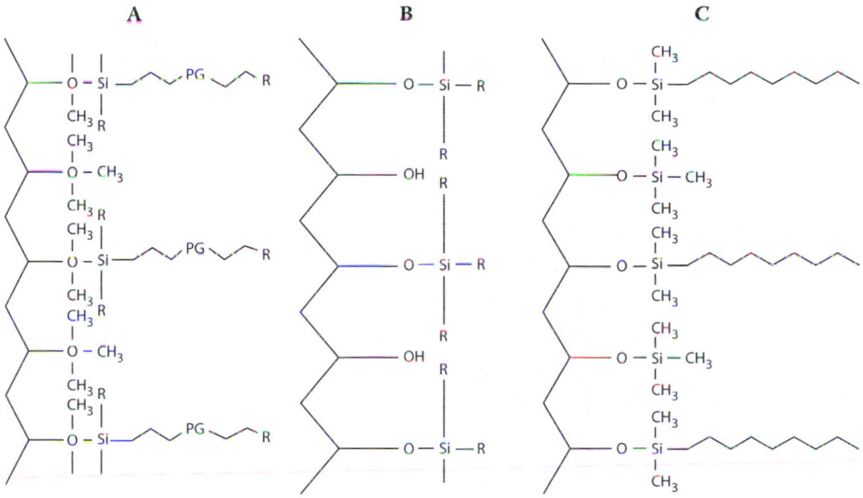

A          B          C

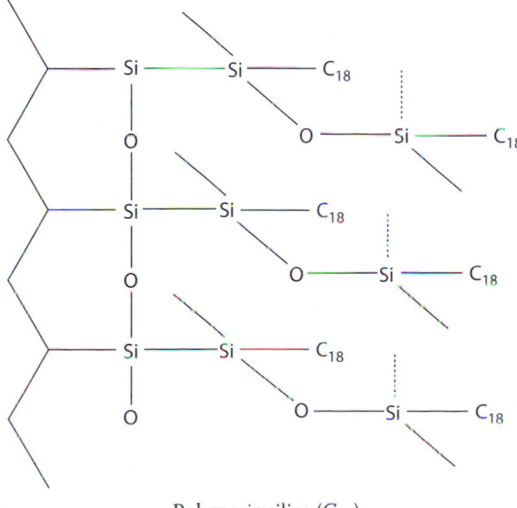

**Diagram A:** shows a modified packing material which utilises bulky silane groups, end capping and a polar embedded functional group

**Diagram B:** shows a modified packing material which utilises bulky silane groups to protect the residual silanol groups

**Diagram C:** shows a modified packing material which utilises end capping

**Figure 6.6** Examples of chemically modified silica packing materials.

Polymeric silica ($C_{18}$)

**Figure 6.7** Example of a chemically modified silica packing material—polymeric bonded silica.

**Table 6.3  Stationary Phase Polarity**

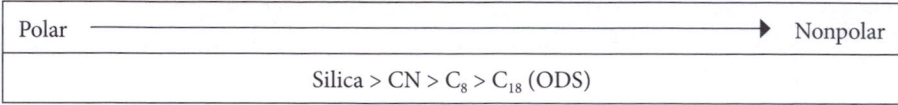

| Polar | ⟶ | Nonpolar |
|---|---|---|
| | Silica > CN > C$_8$ > C$_{18}$ (ODS) | |

manufacturers will have analysed something similar. Column manufacturers provide lots of information on the separation mechanisms associated with a wide range of column chemistries, including ODS, amino (–NH$_2$), and cyano (CN) columns (see Table 6.3). It is worthwhile contacting one of the column manufacturers if you have a particular problem in relation to column selectivity; generally, they are more than happy to help out.

The more sophisticated the column is, the more expensive it is. Generally, it is not advisable to spend large sums of money in the early stages of method development on highly specialised HPLC columns. In the case of amphetamine, a more practical approach might be to use a selection of C$_{18}$ columns that are already in circulation within the laboratory.

**Note:** It is important to mention that HPLC columns do degrade through use (see Chapter 12) and over time; therefore, caution is advised in such cases.

If it is later suspected that a particular column might be suitable for an application, then a new column should be purchased and checked against the early work. Most labs will have a preferred column provider that is usually happy to help out where it can and often will provide columns on a trial basis.

## Carbon Loading and End Capping

The carbon loading is a measure of the carbon content of the silica bonded stationary phase. It is normally quoted as a percentage (some examples are given in Table 6.4). A higher amount of carbon loading has the following advantages:

- It results in a more hydrophobic stationary phase. A more hydrophobic stationary phase results in increased retention and therefore better resolution of hydrophobic analytes. It should be noted, however, that a higher carbon loading does not always result in a better separation. This is of particular significance when dealing with highly nonpolar analytes. In this case, a more hydrophobic stationary phase will have a greater attraction for the nonpolar analyte and may mean that retention times are excessive or that the analyte is retained completely.
- It provides greater column capacity. This allows for a larger sample amount to be injected onto the column, either as a more concentrated

**Table 6.4    Percent Carbon Loading Values for Some ODS (C$_{18}$) RP-HPLC Column Types**

| Column Type | Manufacturer | Carbon Loading (%) | Application |
|---|---|---|---|
| ODS-H | Capital HPLC | 18 | Mixtures of acids and bases |
| ODS-L | Capital HPLC | 15 | Highly polar analytes (e.g., sugars) |
| Zorbax Rx C$_{18}$ | Agilent | 12 | Mixtures of acids and bases |
| Luna C$_{18}$(2) | Phenonenex | 17.5 | All forensic applications |
| Hypersil C$_{18}$ | Thermo Scientific | 10 | pH 1–11; wide range of acidic and basic analytes |
| Nova-Pak C$_{18}$ | Waters | 7 | Mixtures of acids and bases |
| Inertsil ODS-3 | GL Sciences | 15 | Highly basic analytes |
| Inertsil ODS-2 | GL Sciences | 18.5 | Amines and basic pharmaceuticals where hydrogen bonding is an issue |

solution or as a larger sample volume. This might be necessary if there is a low level of anlayte within the sample matrix. A larger column capacity is associated with a larger surface area, which is linked to particle size.

**EXAMPLE 6.6**

You may be required to analyse a sample containing a small amount of amphetamine in the presence of a large amount of caffeine. In order to quantify the amphetamine component, you will need to load a large amount of sample onto the column. This will overload the column with caffeine, leading to peak tailing unless the column capacity is high (see Chapter 10).

- It reduces the number of residual silanol groups, which can cause secondary interactions, especially at low pH values. This is because the silanol groups become charged, resulting in unwanted ion-exchange and hydrogen-bonding reactions taking place between the silanol groups and the analytes of interest. This is another cause of peak tailing and therefore its elimination due to higher carbon loading has a similar benefit to that described previously. Residual silanol groups on the surface of RP-HPLC columns are acidic in nature; pH values range from about 3 to 9.5. According to Forgacs and Cserhati (1997), the acidity will vary depending on the nature of the silanol groups that are present. This is also documented by Herrero-Martinez et al. (2004).

  Because of the wide range of pKa values that have been reported, it would be likely that at almost any mobile phase pH, some of the silanol groups will be charged and some will be uncharged. Let us say that the mobile phase pH is 7. If we use amphetamine as an example, we already know that at pH 7, it will be in the ionised form, which can be represented as R$_3$NH$^+$. The R$_3$NH$^+$ will interact with the residual OH⁻ groups on the packing material surface, resulting in significant peak tailing.

- Higher carbon loading provides greater stability at extremes of pH. It also provides a shield to protect exposed silanol groups from attack by the mobile phase. This is beneficial because it allows the chromatography to take place at more extreme pH levels and is advantageous because it allows adjustment of the pH in the mobile phase so that analytes of interest remain in the non-ionised form. This results in enhanced retention times and hence better separations.

**EXAMPLE 6.7**

Let us consider the analysis of gamma hydroxbutyrate (GHB; see Figure 6.8), which has a pKa of 4.72 (see Figure 6.9). In order to maintain the analyte in a non-ionised state, the system would need to be operated below pH 2, which is outside the range for conventional silica-based columns. With a low carbon load column, operating at this pH level would not be possible; however, the high carbon loading on some columns allows for pH operation at values as low as pH 1. Several manufacturers offer columns of this nature that can be accessed via their Web sites.

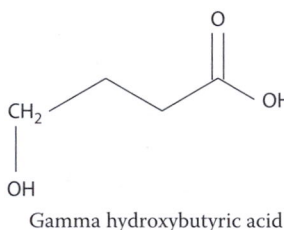

Gamma hydroxybutyric acid

**Figure 6.8** Structure of GHB.

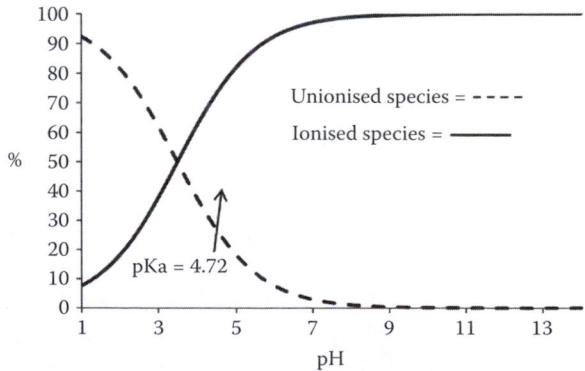

**Figure 6.9** Ionisation of gamma hydroxbutyric acid with change in pH.

## Pore Size

Consideration of the pore size in method development is important when the separation process involves molecules of high molecular weight. Most reversed phase packing materials have a pore size of between about 80 and 120 Å. Some manufacturers quote the pore size in microns ($\mu$m) and some use angstroms (Å). There is a range of pore sizes, depending on the manufacturer and the type of packing material. In forensic science, the pore size is not too great a consideration and, again, the suggestion would be to use what is already in the laboratory.

According to chromatography theory (Chapter 2), analytes interact with the stationary phase as they pass through the column. As part of this process, the analytes move in and out of the pores in the stationary phase beads. For small molecules, this is generally not a problem and the analytes can move freely in and out of the stationary phase pores. If larger molecules are the subject of the investigation, then a larger pore size will be required to accommodate the larger molecule. If a column with too small a pore size is used, there will be little, if any, interaction with the stationary phase and the compounds will pass straight through the column.

## Particle Shape and Size

HPLC packing materials are made up of tiny spherical beads that vary in size from 2 to 50 $\mu$m. It is important that the particle size within any given column lie within a narrow range. This is a complex process and manufacturers' columns do differ; however, well-made columns will have a narrower particle size distribution. It is not uncommon for manufacturers to advertise column packing materials based on the uniformity of the particle size. This is normally demonstrated using a distribution curve.

As we saw in Chapter 2, the smaller the particle size is and the more uniform the packing materials are, the more efficient is the column because of a reduction in the multiple path effect or eddy diffusion. Smaller, more uniform particles promote a more even path through the column and result in sharper, more efficient peak shapes. This is reflected in higher plate height values. More information on this topic can be found in the 'Further Reading' section at the end of this chapter. A smaller particle also means that more particles can be packed into a column of a given size, which in turn means that more of the stationary phase will be present. If this is the case, then the capacity of the column will be increased, allowing more concentrated samples to be injected. This is particularly useful when the method must accommodate a wide range of sample concentrations, such as those that might be encountered in forensic applications.

The downside of smaller particles is that pressure will increase as the particle size decreases. As the number of particles increases within the column, a much higher pressure will be required to push the mobile phase through the system.

## Column Dimensions

Column dimensions include the length and the diameter of the column, each of which has a role to play in the method development strategy. The more time our analytes spend in the column, the greater will be the separation. It follows, then, that if we were to have a longer column, we would expect to see longer retention time for a given mobile phase and flow rate. This means that if we have poor resolution between a particular set of peaks, we can improve it by lengthening the column.

Columns vary from 5 to 25 cm in length. Longer columns tend to be more expensive and use a more mobile phase, which can be a consideration within any lab. Longer retention times also mean longer analysis times, which can also be an issue. Sometimes, instead of increasing the column length, it is better to consider using a different column selectivity, thus affecting the separation at an acceptable and reasonable cost.

Something else to consider in relation to longer columns is that if the sample remains on the column for a longer time, band broadening will increase due to longitudinal diffusion (see Chapter 2). Narrower columns, on the other hand, have less of an effect due to band broadening. Larger diameter columns allow for greater sample concentration and require lower operating pressures. Smaller diameter columns require much higher pressure and thus tend to operate with much lower flow rates. This can mean that the system will need to be modified to accommodate the lower flow rates, which can be as low as 0.1–0.2 mL/min.

## Choosing a Mobile Phase

Most separations in RP-HPLC use either methanol or acetonitrile as the organic modifier in the mobile phase. Downsides are associated with both of these solvents in terms of cost and toxicity and in relation to their UV absorbance at low wavelengths. Different solvents have different polarities and can be categorised according to their polar index (P), which is a measure of the solvent polarity (see Table 6.5). As a rule of thumb, the higher the polarity index is, the more polar is the solvent.

If our analytes are retained on the column for too long, we need to decrease the polarity of the mobile phase by selecting a solvent with a lower polarity index. For example, if we started our separation using methanol and our analytes were retained for too long, we could adjust the elution power

**Table 6.5  Polarity Index Table**

| Solvent | Formula | Polarity Index (P) | UV Cut-off (nm) |
|---|---|---|---|
| Acetone | $C_3H_6O$ | 5.1 | 330 |
| Acetonitrile | $C_2H_3N$ | 6.20 | 210 |
| n-Butanol | $C_4H_{10}O$ | 4.00 | 220 |
| Chloroform | $CHCl_3$ | 4.10 | 245 |
| Cyclohexane | $C_6H_{12}$ | 0.20 | 210 |
| Hexane | $C_6H_{14}$ | 0.10 | 210 |
| Heptane | $C_7H_{16}$ | 0.10 | 200 |
| Methanol | $CH_4O$ | 5.10 | 210 |
| Pentane | $C_5H_{12}$ | 0.00 | 210 |
| 2-Propanol | $C_3H_8O$ | 3.90 | 210 |
| Tetrahydrofuran (THF) | $C_4H_8O$ | 4.00 | 220 |
| Toluene | $C_7H_8$ | 2.40 | 285 |
| Water | $H_2O$ | 10.20 | 200 |

by changing to acetonitrile instead. This would cause our analytes to elute more quickly. Because the processes are complex, we may find that the elution order is changed, and this needs to be addressed.

**EXAMPLE 6.8: NONPOLAR ANALYTES—RP-HPLC**

Nonpolar analytes have an affinity with a nonpolar stationary phase and the more nonpolar the analyte is, the greater will be that affinity. If we were to use only water as the mobile phase, then we would expect that the nonpolar analytes would spend their time in the stationary phase. If we introduce an organic solvent into the mobile phase, then we alter the polarity of the mobile phase and it becomes more nonpolar in nature.

The nonpolar analytes are now more likely to spend some time in the stationary phase and some time in the mobile phase, as discussed in Chapter 2 when we talked about the distribution coefficient ($K$). The larger the proportion of organic solvent in the mobile phase is, the more likely it is that the analytes will spend time there and the distribution coefficient will favour the mobile phase. If this happens, then the analytes will travel at a much faster rate through the column, resulting in a shorter retention time.

The same theory applies not only to the amount of organic solvent in the mobile phase but also to the eluting strength of the solvent. The elution strength can be determined from the polarity index. A low value indicates stronger elution strength. For example, acetonitrile has higher elution strength than methanol. If retention times are long, then the proportion of methanol can be increased or the methanol portion of the mobile phase can be substituted with aetonitrile, which will cause the analytes to elute more quickly.

## Automated Method Development

Method development can be a time-consuming process and requires quite a substantial amount of guesswork. Hundreds of columns are available, with different manufacturers claiming to have the best new packing material on the market. It can be difficult to know where to start. Some of the guesswork can be removed from method development by using automated HPLC development software. Many different software packages are available on the market, some of which include:

POPLC: Bischoff Chromatography
32 Karat™ Software: Beckman
Turbo Method Development: Perkin Elmer
ILS ChromSmart MD: Agilent

Initially, the system is programmed by the analyst to run a single preliminary experiment or a series of preliminary experiments using different solvent compositions with acceptance criteria in relation to resolution and retention time. The software can then interpret the data and determine the next experiment that should be run in order to fulfil the desired outcomes. The results can be quickly evaluated and the most promising conditions isolated for further investigation. This all takes a fraction of the time required for manual method development because the system can be left to run through a series of experiments overnight. The automated system still requires the operator to be knowledgeable in relation to HPLC separation mechanisms.

## Method Development Strategy

1. Find out all that you can about the sample, including structure, solubility, etc.
2. Has anyone else developed a method or your analyte before? It is always easier to adapt someone else's method rather than to start from scratch. A thorough literature review is advised before embarking on the method development journey.
3. Think about the intended use of the method—qualitative, quantitative, or both.
4. Start simple. Use reference standards that mimic all of the analytes of interest that might be included in the sample. Do not forget about the matrix but do not spend too much time on it initially. You will need to ensure that you are able to separate the analytes of interest from interfering matrix components, but this can be done at a later date and should not form part of the initial development.
5. Decide on a detection system and prepare reference standards that you know will give a good signal. Some preliminary experiments are advisable in order to check the detector response prior to running any chromatographic separation. For example, if you choose to use

ultraviolet/visible (UV/Vis) spectrophotometry, then you should prepare a solution of the reference standard to determine the wavelength of maximum absorbance and the $A^{1\%}_{1cm}$ for each analyte of interest. UV/Vis detection systems are the most commonly used in HPLC, but you must make sure that the detector will 'see' all of your analytes of interest (i.e., all of the analytes must have a suitable chromophore).

6. Think about what concentrations you might expect to see in your samples when the method goes 'live'.

7. Choose a column that is compatible with your analyte of interest based on the column selectivity. A 15 cm × 4.6 cm stainless steel column containing 5 μm $C_{18}$ or $C_8$ packing material is a good place to start. Choose a column with good retention characteristics for your types of analytes.

8. Think about controlling the pH of the mobile phase by using a buffer. For example, will your sample be ionised within normal operating parameters (around pH 7)? If the answer is yes, then you may need to consider using a buffer. The choice of buffer will be dependent on the pKa of your sample.

9. The choice of mobile phase must be compatible with the HPLC system and must not contain any strong acids or bases that will lead to degradation of the system or the column packing material. All solvents should be of high quality to avoid some of the pitfalls described in Chapter 10. A 50/50 mixture of methanol and water forms a crude but often effective starting point for neutral analytes and a 50/50 mixture of methanol and aqueous buffer forms a starting point for ionisable analytes. Remember to allow the system time to equilibrate before injecting any sample solutions. Flow rates of around 1 mL/min offer a safe place to start.

10. Set up your HPLC system at the required detector setting with a run time of about 20 minutes and an initial injection volume of 10 μL.

11. Prepare a standard solution and perform multiple injections using the same conditions; at least two injections should be performed at every stage of the development. This ensures that any changes have a genuine effect that is reproducible.

12. Refine the method manually by changing one parameter at a time. Keep performing additional changes until the required separation, retention, and peak shape are achieved. This is not a fast process; it can take several weeks, depending on the complexity and nature of the samples.

13. Finally and most importantly: *do not be put off by early failure.* You will not achieve the perfect method at the first attempt. Interpret the chromatograms at each stage and work out what processes are taking place and adjust accordingly. For example, you are trying to

develop a method for a number of weak bases and your initial retention times are approximately 15–20 minutes. Clearly, this is too long,
so you have a number of choices:

a. Decrease the column length, which decreases the retention time.
b. Increase the flow rate, which decreases the retention time but
   uses a more mobile phase.
c. Change the mobile phase. Move to a more nonpolar mobile phase
   by increasing the percentage of the organic modifier. This can
   result in reduced resolution and can have an impact on the elution characteristics, so beware!
d. Change the column. Move to a less retentive packing material
   such as a $C_8$ instead of a $C_{18}$. Again, there will be an impact on the
   resolution and elution characteristics.
e. Adjust the pH of the mobile phase in order to ionise the analytes.
   With a 50/50 methanol/water mix, this has probably already
   happened, but you might want to consider controlling the pH to
   ensure consistent retention times.

This procedure can be simplified with the aid of the selection scheme in
Figure 6.10.

## Method Development Worked Example

To highlight the method development plan, we will look at a series of preservatives known as parabens. These preservatives are used in the cosmetic,
pharmaceutical, and food industries and are effective antifungal and antimicrobial agents. Antifungal and antimicrobial activity increases with
increasing chain length. Methyl and propyl parabens can be added directly
to foodstuffs in most countries, and ethyl paraben is permitted in some
countries. They can be incorporated into foods at a maximum level of 0.1%.
Parabens do occur naturally, but those used in industry are synthetic analytes produced by the esterification of parahydroxbenzoic acid with a suitable
alcohol. For example, methanol would be used in the manufacture of methyl
parahydroxybenzoate. We have chosen to look at a subsection of the paraben
series that includes

- methyl parahydroxybenzoate
- ethyl parahydroxybenzoate
- propyl parahydroxybenzoate

Although parabens have been considered safe for many years, an increasing body of knowledge has challenged this assumption. As a result, sodium

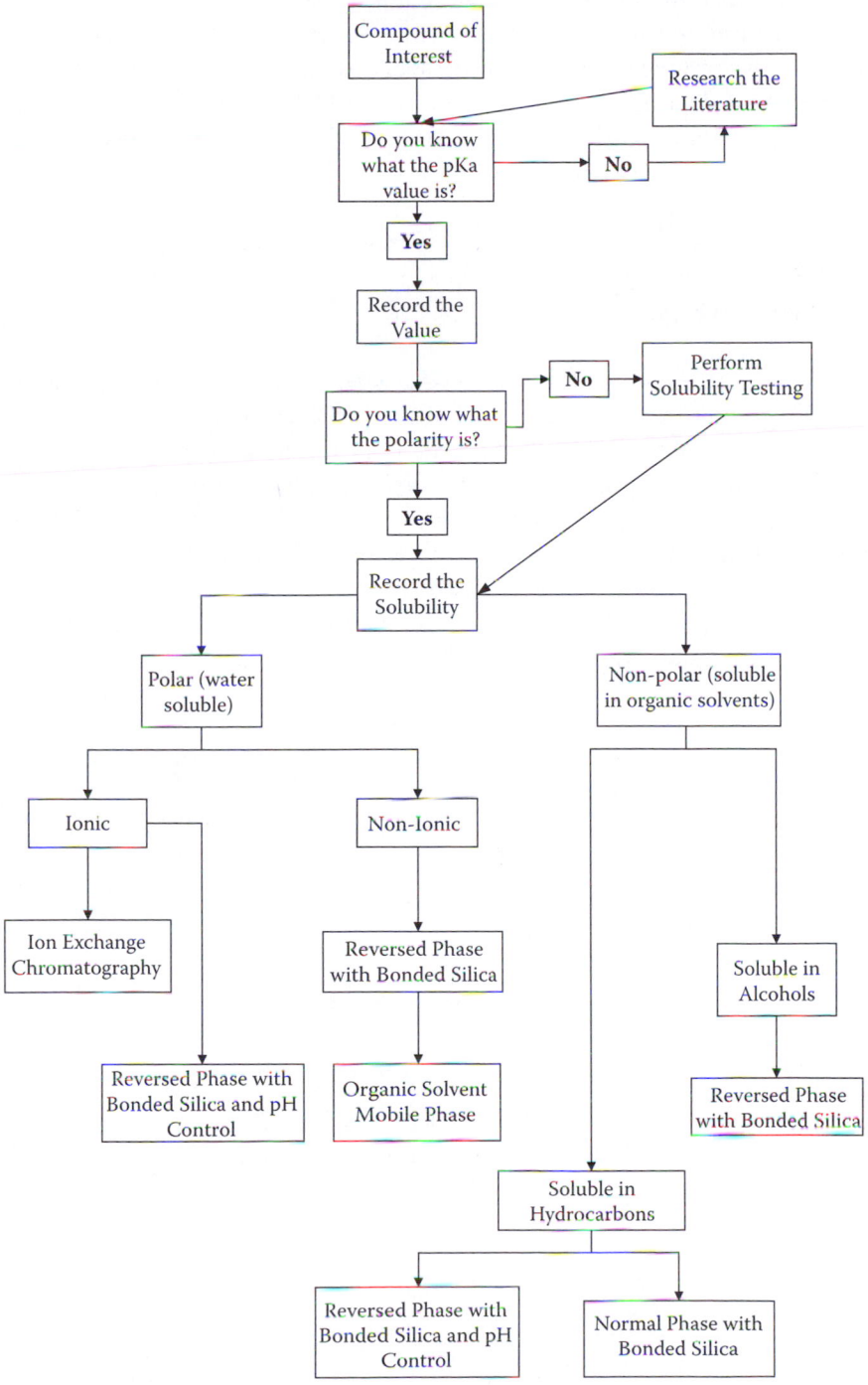

**Figure 6.10** Column selection scheme for HPLC analysis.

propyl paraben has been withdrawn under EU legislation. Parabens are not often encountered in a forensic science environment, but we have chosen them as an example to illustrate this aspect of method development because they are well documented in the literature on HPLC methods. Thus, we can use them to highlight the principles already discussed. As always, the three things that we need to consider are

- the analytes
- the column
- the mobile phase

We know the names of the analytes, so we can start by taking a look at their chemical structure (see Figures 6.11, 6.12, and 6.13). One of the first things that gives us a hint about the molecule is in the name of the alkyl moiety, which tells us that we are looking at a chemical series with increasingly long alkyl chains: methyl ($-CH_3$), ethyl ($-CH_2CH_3$), and propyl ($-CH_2CH_2CH_3$). The chemical structure will give us information about the molecular weight, the polarity, and possible acid or base properties.

We already know from our column chemistry that increasing the alkyl chain length increases the hydrophobicity (or makes the column more

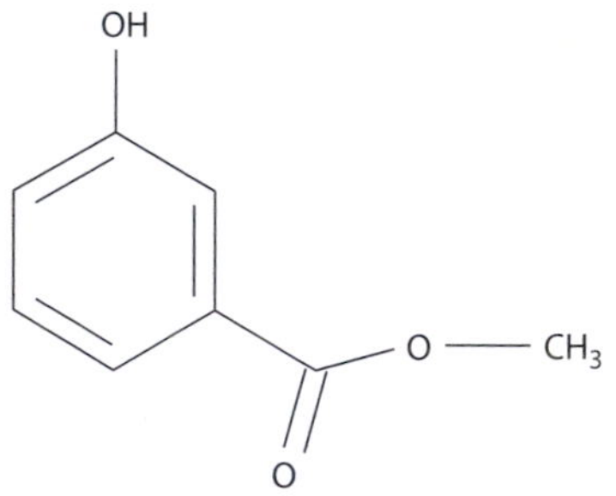

Methyl Paraben pKa = 8.4

**Figure 6.11** Methyl paraben.

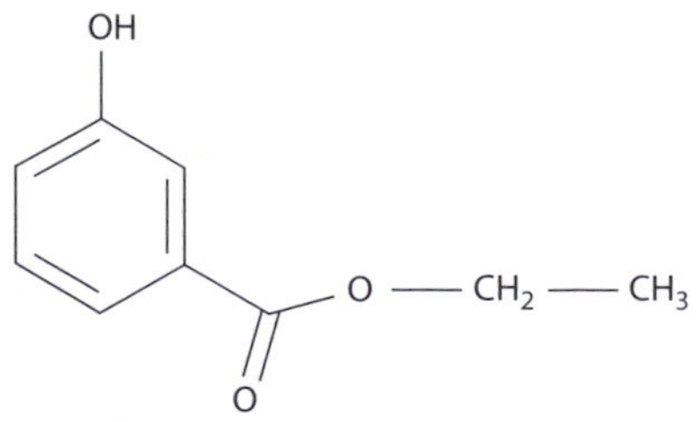

Ethyl Paraben pKa = 8.3

**Figure 6.12** Ethyl paraben.

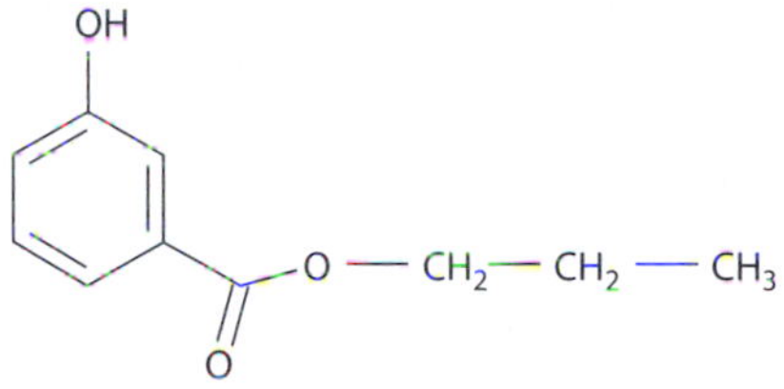

Propyl Paraben pKa = 8.4

**Figure 6.13** Propyl paraben.

**Table 6.6  Solubility of Selected Parabens**

| | Solubility (Solvent) | | |
| Analyte | Water | Ethanol | Chloroform |
| --- | --- | --- | --- |
| Methyl paraben | 1 in 400 | 1 in 3 | 1 in 40 |
| Ethyl paraben | 1 in 1500 | 1 in 2 | 1 in 10 |
| Propyl paraben | 1 in 2500 | 1 in 1.5 | 1 in 4 |

nonpolar) and the same concept also applies to our paraben analytes. Thus, this tells us that as we progress through the series from methyl to propyl parahydroxybenzoate, the molecules become more nonpolar in nature because of the increase in the number of C–C and C–H bonds in the molecule. The next thing to examine is the solubility information. This will give us additional information in relation to the polarity of our analytes. Solubility information is provided in Table 6.6.

If we look at the solvents in the table, we know that water is more polar than ethanol, which in turn is much more polar that chloroform (see Table 6.4). We have already established that our parabens become more nonpolar with increasing alkyl chain length. From this we would expect to see that propyl paraben would be more soluble in chloroform than methyl paraben and that methyl paraben would be more soluble in water than propyl paraben. This can be verified from the data in Table 6.5. From our solubility data, we can see that all of the parabens have a greater level of solubility in relatively more nonpolar solvents (relative to water). We can deduce from this that our analytes are less polar than water but not as nonpolar as chloroform.

If we take a closer look at the functional groups present in the molecule (see Figure 6.14), we can see that a nonpolar element is associated with the alkyl chain (C–C and C–H bonds), a slightly less nonpolar element associated with the aromatic ring, a slightly more polar element associated with the ester group, and a more polar group associated with the phenol group (–OH). Overall, the molecule is slightly nonpolar in nature. We know that the analytes are compatible with the types of solvent that we might expect to

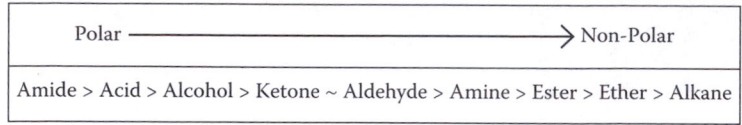

**Figure 6.14** Functional group polarity.

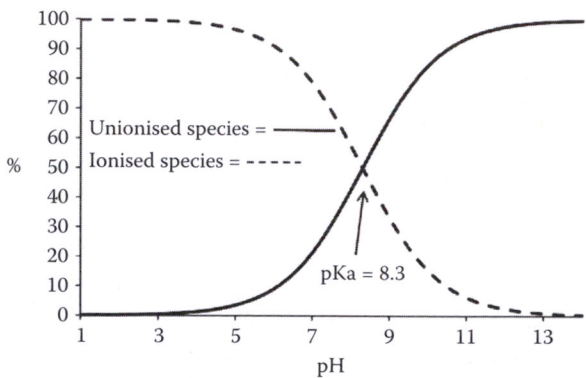

**Figure 6.15** Ionisation of parabens with changing pH.

use in RP-HPLC and that the analytes are slightly nonpolar in nature. Taking all of this into account, it would make sense to start with RP-HPLC using a $C_{18}$ column and an aqueous or organic mobile phase.

Next, we need to look at the pKa values and assess the effect of a change in pH on the molecules. If we take methyl paraben as an example, we would expect to find ionisation at the –OH group attached to the aromatic ring (see Figure 6.11). Because these analytes are acidic in nature (if you need to refresh your memory, go back to Chapter 2), we would expect the non-ionised form to dominate at low pH and the ionised form to dominate at high pH (see Figure 6.15).

In summary, we have slightly nonpolar acidic analytes. Reviewing all of the information, we might want to start with a nonpolar column and an aqueous or organic mobile phase buffered to approximately pH 3. Before we go down the buffer route, let us see what happens if we use a simple methanol and water mix. In order to compare conditions during the development process (following experiments 1–4), we have calculated several parameters (reported in Table 6.7). The equations used to calculate the different parameters are given in Chapter 2.

## Experiment 1

Let us try a $C_{18}$ column with methanol/water (50/50) as the mobile phase. According to Figure 6.15, at pH 6–7 we would expect to see approximately

**Table 6.7   Experimental Data**

| | Experiment No. | | | |
|---|---|---|---|---|
| | 1 | 2 | 3 | 4 |
| $t_R$—Methyl paraben (min) | 3.89 | 4.00 | 2.38 | 2.68 |
| $t_R$—Ethyl paraben (min) | 5.97 | 6.15 | 2.83 | 3.38 |
| $t_R$—Propyl paraben (min) | 10.57 | 10.88 | 3.68 | 4.75 |
| Resolution (peak 1:2) | 5.7 | 5.8 | 1.3 | 3.6 |
| Resolution (peak 2:3) | 8.9 | 8.9 | 2.9 | 5.0 |
| $k'$ | 1.4 | 1.5 | 0.5 | 0.7 |
| $N$ (peak 3) | 5053 | 1087 | 3828 | 4325 |
| Peak assymetry (peak 3) | 1.08 | 1.08 | 1.37 | 1.18 |

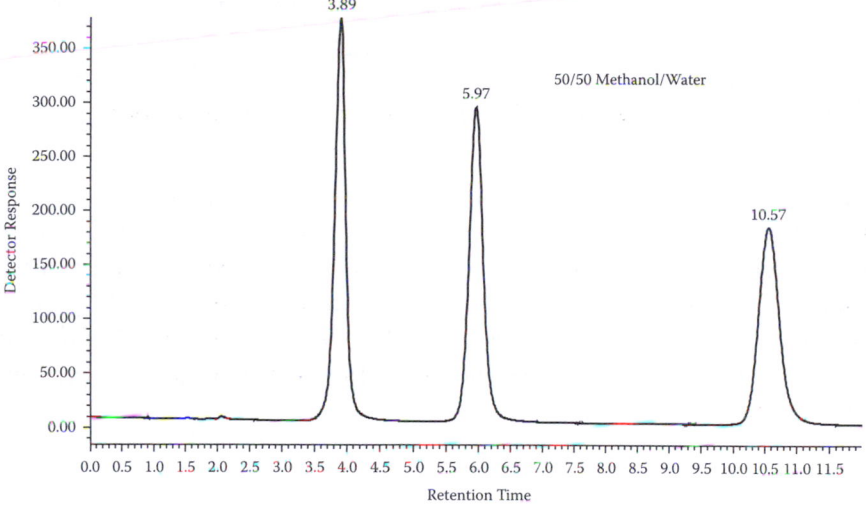

**Figure 6.16** Methyl, ethyl, and propyl paraben—50/50 methanol water, column—GraceSmart® RP-C$_{18}$ 150 mm × 4.6 mm, 5 μm.

10% of parabens in the ionised form. The results of this analysis are shown in Figure 6.16. We can see from the chromatogram that we have good separation of the three analytes. However, the retention times are longer than necessary (see Table 6.6). In Chapter 2, we discussed the ideal values for these parameters as being $k' > 1$ and $\leq 10$ and $R \geq 1.5$.

## Experiment 2

Before we rush ahead and increase the percentage of organic modifier, let us take a look at what would happen if we maintained the proportions within the mobile phase but adjusted the pH of the aqueous phase to pH 3. Referring

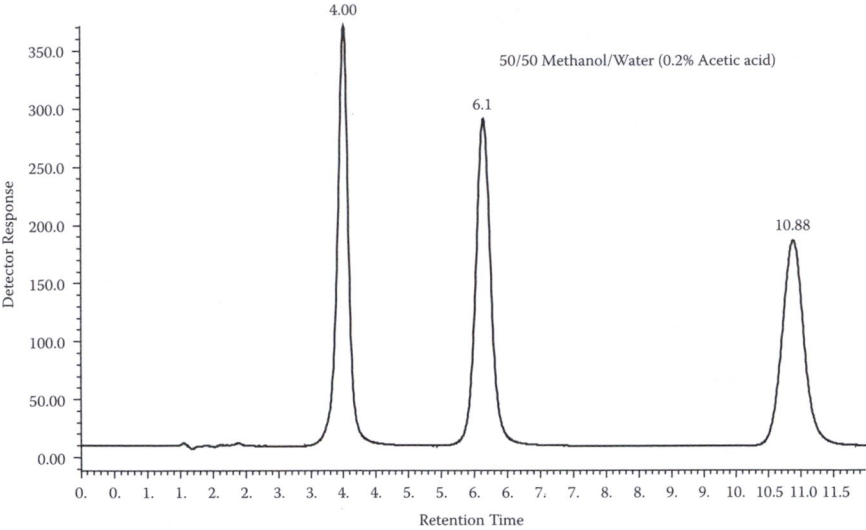

**Figure 6.17** Methyl, ethyl, and propyl paraben—50/50 methanol water (0.2% acetic acid), column—GraceSmart RP-C$_{18}$ 150 mm × 4.6 mm, 5 µm.

back to Figure 6.15, we would expect that all of the parabens would be in the non-ionised form. Let us see what happens if we inject our sample mixture under these conditions (see Figure 6.17). We can see that there is a very small change in retention time. Retention times have increased slightly, as we would expect. All of the paraben analytes are non-ionised (R–OH) and are considered to be in the most nonpolar physical state—hence the longer retention times.

## Experiment 3

We need to reduce the retention times for all of the analytes and, according to our theory, we can do this by increasing the proportion of organic modifier in our mobile phase. Let us see what happens when we change the mobile phase composition to 60/40 methanol/water (see Figure 6.18). Here we can see that the retention times are shorter, but we have failed to meet our acceptance criteria for $k'$ and resolution (see Table 6.7).

## Experiment 4

We can adjust the mobile phase composition again, so let us try 55/45 methanol/water (0.2% acetic acid). We should expect to see slightly longer retention times (see Figure 6.19). Our chromatogram shows longer retention time and once again we have met our acceptance criteria for $k'$ and resolution. This method would now be acceptable in terms of our acceptance

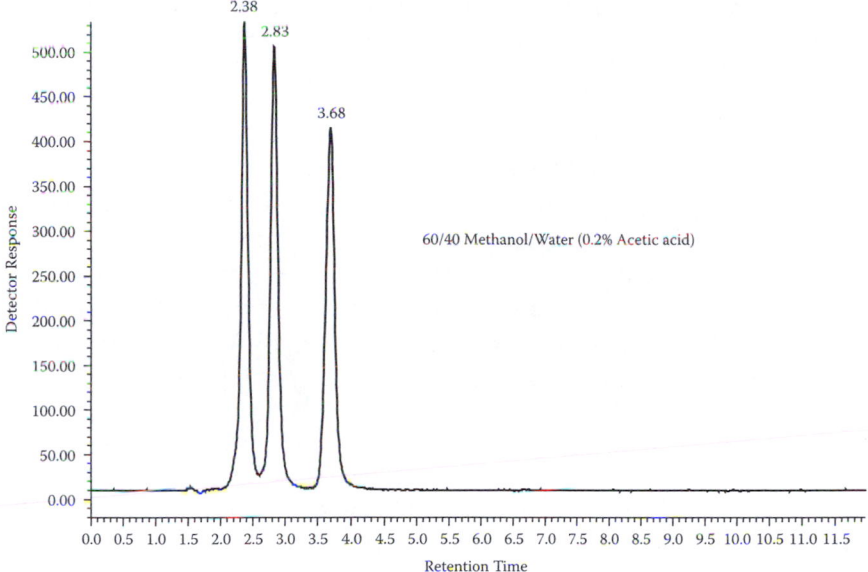

**Figure 6.18** Methyl, ethyl, and propyl paraben—60/40 methanol water (0.2% acetic acid), column—GraceSmart RP-C$_{18}$ 150 mm × 4.6 mm, 5 μm.

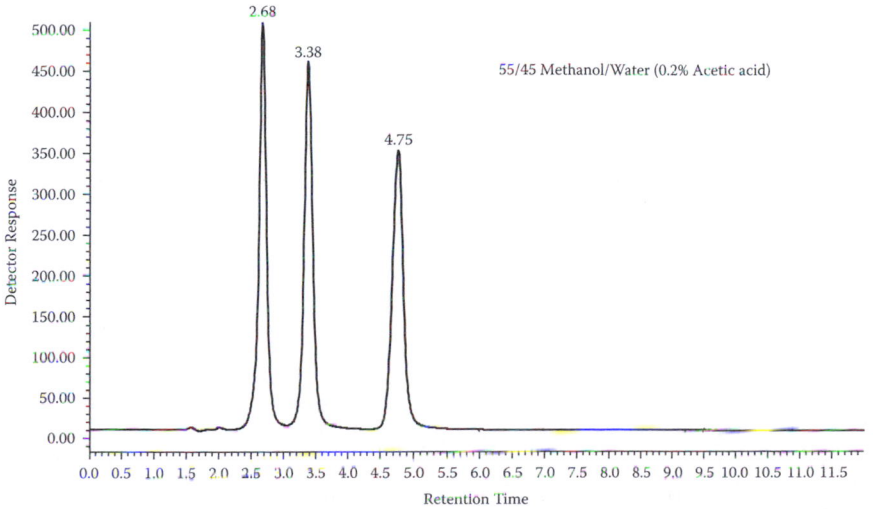

**Figure 6.19** Methyl, ethyl, and propyl paraben—55/45 methanol water (0.2% acetic acid), column—GraceSmart RP-C$_{18}$ 150 mm × 4.6 mm, 5 μm.

criteria and we could go ahead and validate the method for routine use in our laboratory.

## Experiment 5

Let us consider what would happen if we changed the column length. An increase in column length means more packing material in the column, so the analytes would have a longer interaction and hence a longer retention time. Let us see what happens if we change the column length to 250 mm. We will maintain the same flow rate, mobile phase (55/45 methanol/water, 0.2% acetic acid), internal diameter, and particle size. Here we can see that we have increased the retention times and the resolution (see Figure 6.20).

## Experiment 6

Think about what might happen if we used a $C_{18}$ column but changed the manufacturer. From our theory, we know that manufacturers' packing materials differ. These differences can be due to pore size, carbon loading, end capping, etc. We would expect to see differences in retention time and separation when we use different manufacturers' packing materials. Let us see what happens if we use two columns, each containing $C_{18}$ packing material but supplied by two different manufacturers. We will maintain the same flow rate, mobile phase (55/45 methanol/water, 0.2% acetic acid), column dimensions, and particle size. Here we can see that we have increased the retention times and the resolution for some of the peaks (see Figure 6.21).

## Experiment 7

Maybe we could change the packing material. From our theory, we know that there are differences in retention depending on the type of packing material used. We know that, for nonpolar analytes, we would expect to see a reduction in retention time if we substitute a phenyl column for a $C_{18}$ column. Let us see

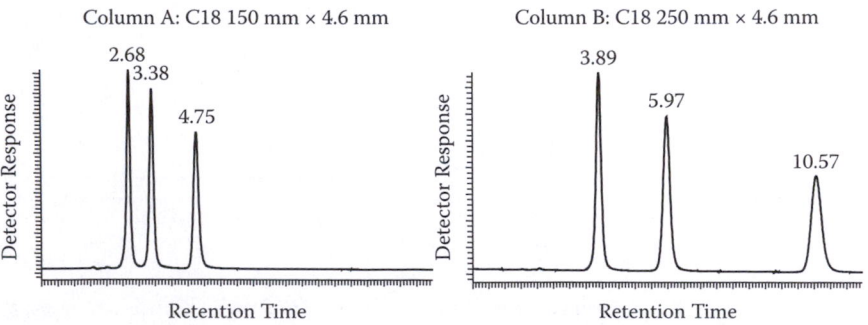

**Figure 6.20** Effect of column length on retention time.

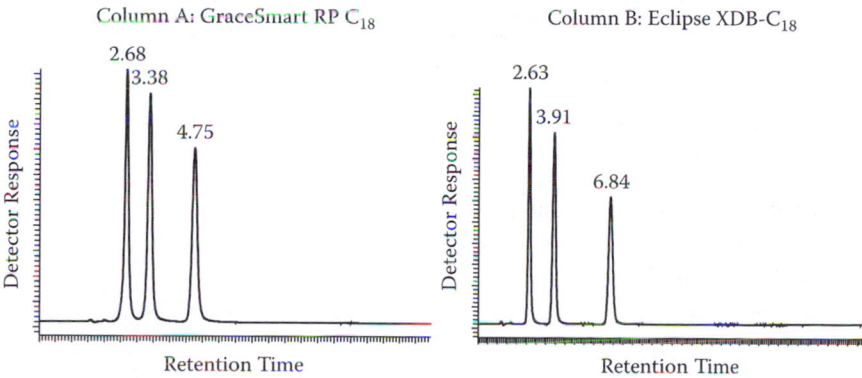

**Figure 6.21** Effect of column manufacturer on retention time.

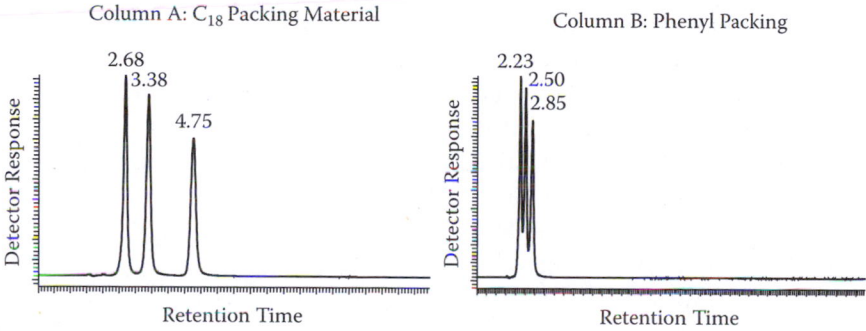

**Figure 6.22** Effect of packing material type on retention time.

what takes place if we use two columns containing different packing material. We will maintain the same flow rate, mobile phase (55/45 methanol/water, 0.2% acetic acid), column dimensions, and particle size. Here we can see differences in the retention times and in the resolution (see Figure 6.22).

---

**KEY POINT SUMMARY**

**Sample Composition**

- Matrix: will there be a requirement to separate the analyte of interest from the matrix?
- Molecular structure/molecular weight.
- pKa: important when choosing a buffer.
- Detection characteristics: does the analyte of interest have a chromophore?
- Solubility: in what is the analyte of interest soluble (see Figure 6.23)?

*(continued on next page)*

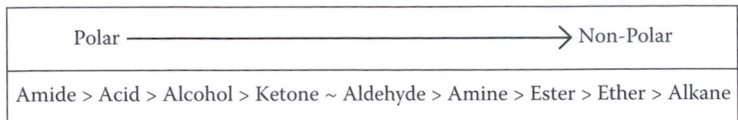

| Polar ─────────────────────────→ Non-Polar |
| --- |
| Amide > Acid > Alcohol > Ketone ~ Aldehyde > Amine > Ester > Ether > Alkane |

**Figure 6.23** Functional group polarity.

## Choosing a Column

- Separation mechanism: what mechanism will be best suited for your analyte?
- Column chemistry: justify your choice.
- Carbon loading and end capping: will there be interaction with residual silanol groups or do you need to be able to inject highly concentrated solutions, possibly at high pH?
- Pore size: small molecule or larger molecule?
- Particle shape and size: is there a need for greater column efficiencies and faster chromatography?
- Column dimensions: strike a balance between resolution and retention time (see Figure 6.24).

| HPLC Mode | |
| --- | --- |
| Rp-HPLC | Nonpolar stationary phase/polar mobile phase |
| Normal phase | Polar stationary phase/nonpolar mobile phase |
| Ion-exchange | Strong acids and bases |

**Figure 6.24** Column chemistry.

## Choosing a Mobile Phase

- Effect of pH: variation in retention times can occur with non-buffered systems.
- Addition of organic modifier: strike a balance of cost, retention time, and resolution (see Figure 6.25).

| pH | Acidic Analytes | Basic Analytes |
| --- | --- | --- |
| Acidic | Non-ionised | Ionised |
| Basic | Ionised | Non-ionised |
| **Organic Modifies** | **Nonpolar Analytes** | **Polar Analytes** |
| Nonpolar | Shorter $t_R$ | Longer $t_R$ |
| Polar | Longer $t_R$ | Shorter $t_R$ |

**Figure 6.25** pH and polarity.

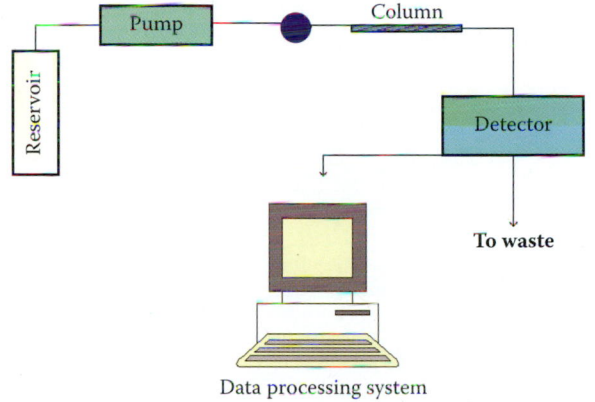

**Figure 1.1** Block diagram of HPLC instrument.

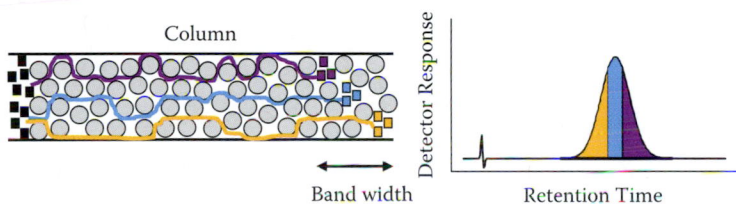

**Figure 2.6** Eddy diffusion.

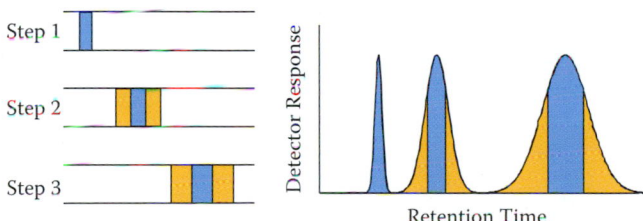

**Figure 2.7** Longitudinal diffusion.

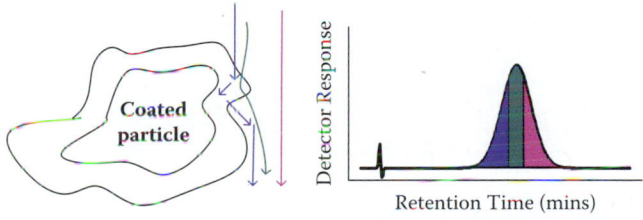

**Figure 2.8** Resistance to mass transfer.

(a)                                                    (b)

**Figure 3.8** Liquid extraction. (a) Two distinct, separate layers; the lower (colourless) layer is dichloromethane and the top layer is the aqueous layer. (b) The formation of an emulsion between the two layers.

**Figure 3.10** Solid phase extraction cartridges of varying size.

Mobile Phase Water pH 8/ACN (30%)
H2O/ACN
H2O/ACN
H2O/ACN
H2O/ACN
Acid+
Acid+
Acid+
H2O/ACN
Acid+
C18
C18
C18
C18
C18
C18
C18
C18
H2O/ACN
Base
Base
Base
Base
OH
Si

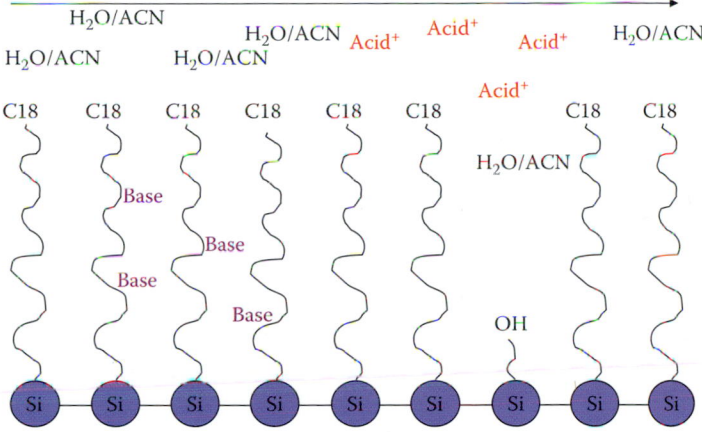

**Figure 4.2** Example of RP-MPLC separation.

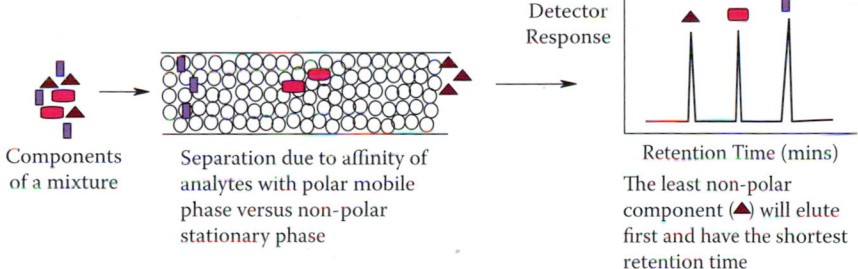

Detector
Response
Components
of a mixture
Separation due to affinity of
analytes with polar mobile
phase versus non-polar
stationary phase
Retention Time (mins)
The least non-polar
component (▲) will elute
first and have the shortest
retention time

**Figure 4.4** Separation in RP-chromatography.

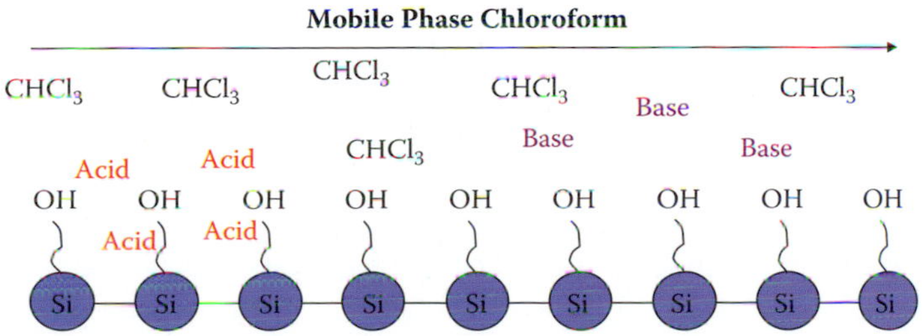

Mobile Phase Chloroform
CHCl3
CHCl3
CHCl3
CHCl3
Base
CHCl3
Base
CHCl3
Base
Acid
Acid
OH
Acid
Acid
Si

**Figure 4.5** Examples of NP-HPLC separation.

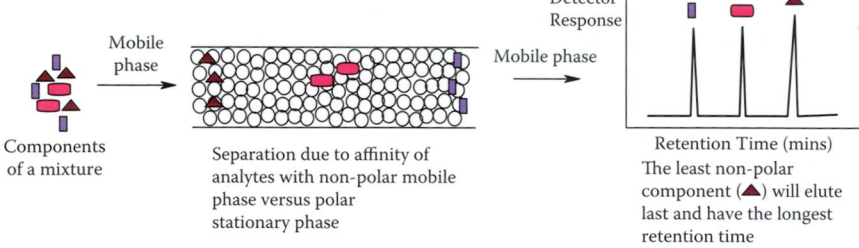

**Figure 4.6** Separation in NP-chromatography.

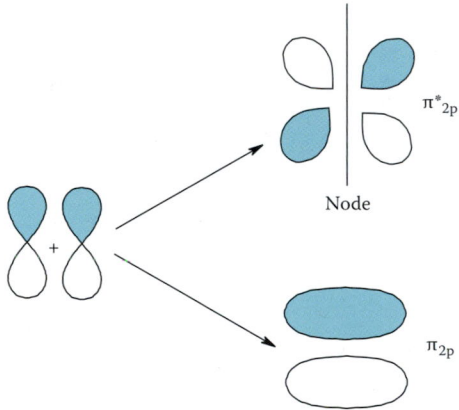

**Figure 5.9** p-Orbitals interacting to form two bonding and two antibonding π-orbitals.

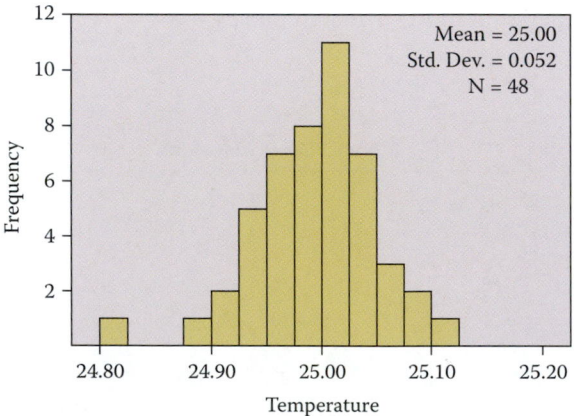

**Figure 8.2** Normal distribution of temperature data set.

# References

Forgacs, E., and T. Cserhati, T. 1997. *Molecular basis of chromatographic separation.* Boca Raton, FL: CRC Press.

Herrero-Martinez, J. M., A. Mendez, E. Bosch, and M. Roses. 2004. Characterisation of the acidity of residual silanol groups in microparticulate and monolithic reversed-phase columns. *Journal of Chromatography A* 1060: 135–145.

# Further Reading

## Books/Online Publications

Agilent Technologies. 2009. *LC method development & transfer: Frequently asked questions* (www.home.agilent.com).

Ahuja, S., and H. Rasmussen. 2007. *HPLC method development for pharmaceuticals.* New York: Elsevier, Academic Press.

Dong, M. W. 2006. *Modern HPLC for practicing scientists.* New York: John Wiley & Sons.

Kazakevich, Y., and R. LoBrutto. 2007. *HPLC for pharmaceutical scientists.* New York: John Wiley & Sons.

Phenomenex. 2009. *Analysis of basic drugs by LC/MS/MS with high pH mobile phases* (www.phenomenex.com).

Sadek, P. C. 2002. *The HPLC solvent guide.* New York: Wiley-Interscience.

SGE International Pty. Ltd. 2001. *How do I develop an HPLC method?* (www.sge.com).

Snyder, L. R., J. J. Kirkland, and J. L. Glajch. 1997. *Practical HPLC method development.* New York: John Wiley & Sons.

Waters Corporation. 2009. *Forensic toxicology* (www.waters.com).

Waters Corporation. 2009. *Method development/validation* (www.waters.com).

## Journal Articles

Baiocchi, C., M. C. Brussino, E. Pramauro, A. B. Prevot, L. Palmisano, and G. Marci. 2002. Characterisation of methyl orange and its photocatalytic degradation products by HPLC/UV-VIS diode array and atmospheric pressure ionisation quadrupole ion trap mass spectrometry. *International Journal of Mass Spectrometry* 214: 247–256.

Boonkerd, S., M. R. Detaevernier, J. Vindevogle, and Y. Michotte. 1996. Migration behaviour of benzodiazepines in micellar electrokinetic chromatography. *Journal of Chromatography A* 756: 279–286.

Coopman, V., M. De Leeuw, J. Cordonnier, and W. Jacobs. Suicidal death after injection of a castor bean extract (*Ricinus communis L.*). *Forensic Science International* 189: e13–e20.

Joyce, J. R., D. G. Sanger, and I. J. Humphries. 1982. The use of HPLC for the discrimination of a range of Dylon home-dyeing products and its potential use in the comparison of illicit tablets. *Journal of the Forensic Science Society* 22 (4): 337–341.

LoBrutto, R., A. Jones, Y. V. Kazakevich, and H. M. McNair. 2001. Effect of the eluent pH and acidic modifiers in high-performance liquid chromatography retention of basic analytes. *Journal of Chromatography A* 913: 173–187.

Lopes Marques, R. M., P. J. Schoenmakers, C. B. Lucasius, and L. Buyden. 1993. Modelling chromatographic behaviour as a function of pH and solvent composition in RPLC. *Chromatographia* 36: 83–95.

Maralikova, B., and W. Weinmann. 2004. Confirmatory analysis for drugs of abuse in plasma and urine by high-performance liquid chromatography—Tandem mass spectrometry with respect to criteria for compound identification. *Journal of Chromatography B* 811: 21–30.

Moeller, M., S. Steinmeyer, and T. Kraemer. 1998. Determination of drugs of abuse in blood. *Journal of Chromatography B* 713: 91–109.

Steenkamp, P. A., F. R. van Heerden, and B. E. van Wyck. 2002. Accidental fatal poisoning by *Nicotiana glauca*: Identification of anabasine by high-performance liquid chromatography/photodiode array/mass spectrometry. *Forensic Science International* 127 (3): 208.

Steentoft, A., and K. Linnet. 2009. Blood concentrations of clonazepam and 7-aminoclonazepam in forensic cases in Denmark for the period 2002–2007. *Forensic Science International* 184 (1–3): 74–79.

Wang, X., J. Yu, M. Xie, Y. Yoa, and J. Han. 2008. Identification and dating of the fountain pen ink entries on documents by ion-pairing high-performance liquid chromatography. *Forensic Science International* 180: 43–49.

Yanagihara, Y., K. Yasukawa, U. Tamura, T. Uchida, and K. Noguchi. 1987. Characteristics of a new HPLC column packed with octadecyl-bonded polymer gel. *Chromatograpia* 24 (1): 701–704.

## Journals (General)

*Analytical Abstracts,* Royal Society of Chemistry (Great Britain), Society for Analytical Chemistry, Chemical Society (Great Britain)

*Forensic Science International,* Elsevier Seqoia

*Journal of the Canadian Society of Forensic Science,* Canadian Society of Forensic Science

*Journal of Liquid Chromatography,* Dekker

*Science & Justice: Journal of the Forensic Science Society,* Forensic Science Society

# System Suitability

7

## Introduction

One way of detecting that something is not right with a chromatographic system is a failure to meet the system suitability criteria limits. All validated analytical methods should include a set of system suitability tests that assess the critical aspects of the methodology. Limits should be applied to these tests, and the tests should be applied to the method at the start of each run ahead of any samples. System suitability runs should be performed using solutions of appropriate mixtures of reference standard material. The data from the system suitability measurements should be assessed on a pass/fail basis against the limits prior to preparing and running any sample material.

Due to often limited sample size, it is an absolute 'must' not to waste sample material by preparing it before the HPLC method is considered suitable for use. Problems encountered at this stage may take some time to diagnose and repair (see Chapter 10) and samples can and do degrade with time (see Chapter 6). System suitability parameters can usually be defined in two ways: the system performance and the chromatographic suitability.

## System Suitability

System suitability does not assess whether any mistakes have been made in sample preparation. The Centre for Drug Evaluation and Research (CDER) document entitled *Validation of Chromatographic Methods* defines the parameters associated with system suitability testing as discussed in the following sections.

### System Performance

#### *Repeatability of Injection (Precision)*
Repeatability assesses the performance of the HPLC system to ensure that there are no issues relating to the injector, the temperature, and the flow rate. It does not take into account any variance that may occur during or at the end of the run. In order to assess the system at these times, it is necessary to

repeat the test during and after the HPLC sequence of samples. Repeatability is normally assessed by performing a series of five or six injections of the 100% target reference standard solution. The relative standard deviation (RSD) of the peak area information is assessed against a predetermined limit. *A limit of RSD ≥ 1% is considered acceptable for most forensic applications.*

## Chromatographic Suitability

### Capacity Factor

The capacity factor ($k'$) is a measure of retention time of the peak of interest ($t_R$) proportional to the column void time ($t_M$)—that is, the retention time for unretained material. The capacity factor is given by the following equation:

$$k' = (t_R - t_M)/t_M$$

*A limit of k' > 2 is considered acceptable for most forensic applications.*

### Relative Retention

The relative retention time is a measure of the retention time of two independent peaks relative to one another and is represented by the following equation:

$$A = k'_1/k'_2$$

### Resolution

Resolution is a measure of how well two peaks are separated and it is a necessary parameter when quantitative measurement is conducted to ensure that accurate assessment of peak area is achieved. It is also a useful parameter to measure where there may be interfering peaks eluting close to the peaks of interest. This is of particular interest in forensic applications where the exact sample composition is unknown. Resolution is measured using the following equation:

$$R = 2(t_{R2} - t_{R1})/(w_1 + w_2)$$

*A limit of R > 1.5 between peaks of interest is considered acceptable for most forensic applications.*

### Tailing Factor

As discussed in Chapter 10, columns begin to degrade with time and, as a result, peak tailing can become an issue. The greater the amount of peak tailing, the more difficult it becomes to measure the peak area accurately. This happens because it becomes more difficult for the integrator to distinguish where and when the peak ends. This can be overcome, to a certain extent, by adjusting the peak width setting in the integration parameters. Increasing the peak width will result in broader peaks. The tailing factor can be measured using the following equation:

$$T = (w_{A5\%} + w_{B5\%})/2w_{A5\%}$$

*A limit of* T $\leq$ 2 *is considered acceptable for most forensic applications.*

### Theoretical Plate Number (N)

The theoretical plate is a hypothetical measurement of the efficiency of the column. The theoretical plate number should be determined upon first use of the column and the value monitored throughout the column lifetime. Using the last peak in the chromatogram, the measurement of $N$ should be determined; it can be calculated using the following equation:

$$N = 5.54 \, (t_R/w_{1/2})^2$$

*A limit of* N > 2000 *is considered acceptable for most forensic applications; however, a database of information should be generated for each column type.*

The Centre for Drug Evaluation and Research (1994) states:

> System suitability testing is essential for the assurance of quality performance of the chromatographic system. The amount of testing required will depend on the purpose of the test method...each method submitted for validation should include an appropriate number of system suitability tests defining the necessary characteristics of that system.

Some forensic laboratories will run what is known as a QC (quality control) sample ahead of the analytical sequence to assess whether all of the system suitability criteria are met. The QC sample is a simulated sample of known concentration and is prepared independently by one or several members of staff within the laboratory. Generally, the QC sample is run and assessed against the system suitability criteria before the start of the analytical sequence and again at the end. The concentration of the sample is calculated and assessed against the true value. A limit is set and this must be met in order for the data to be considered valid.

When an analytical sequence is run, it is important that reference standards are inserted throughout the run in order to assess the data quantitatively as well as to monitor system performance. All of the system suitability tests need not be applied to all of the data generated at all stages. However, it is important to inspect the chromatograms for any anomalies.

## Reference

Centre for Drug Evaluation and Research (CDER). 1994. *Validation of chromatographic methods* (www.fda.gov).

# Qualification, Validation, and Verification

# 8

## Introduction

It is important to provide your customer with reliable results that are fit for purpose. As forensic scientists, we look to do this by performing instrument qualifications (calibrations against verified reference materials) and method validations (using verified reference standards), and by ensuring that we apply rigorous system suitability (see Chapter 7) and quality procedures into our methodologies. In relation to HPLC, qualification relates to an assessment of the instrumentation (pump, injector, column compartment, detector) that is used to determine the measured result, whereas validation relates to the applications and methods that are used to generate the results.

Can you imagine going to court with results that you are not completely sure about in terms of validity and reliability? Imagine the situation where you have been asked to attend the coroner's court in relation to a suspected drug overdose: You are in the witness box and you are asked to confirm that your results indicate that John Doe may have died from an overdose of heroin. His family is in court and your answer is, 'I think so, but I can't really be sure it was heroin and I can't be sure about the amount exactly'. Clearly, this answer is not acceptable and it would not take many appearances at court of this type for you to lose any credibility as an expert witness and to bring your laboratory into disrepute. In order to avoid these types of situations, it is normal for forensic laboratories to implement a set of principles that ensure that results are valid, reliable, and repeatable. This forms part of the company's overarching quality system (see Chapter 9). One of the ways in which a laboratory can ensure that measurement is valid and fit for purpose is to adopt the valid analytical measurement (VAM) system.

## Valid Analytical Measurement

The VAM system was introduced and developed by LGC in the 1980s and describes a set of six principles designed to promote best practice and provide valid data to customers through quality programmes. The VAM principles are available through the National Measurement System Chemical and

Biological Metrology Web site. Not every laboratory calls its quality system a VAM system, but all credible laboratories will have a quality system in place. The six principles as described in VAM are discussed next.

### Principle 1: Analytical Measurement Should Be Made to Satisfy an Agreed upon Requirement

This principle is about understanding the problem and ensuring that the customer's needs are met. There seems no point in analysing a sample for cocaine when a heroin analysis has been requested. Forensic science is not clear-cut by any means and it may be necessary to readdress the requirements of an analysis on completion of the initial results. An example of this might be when the circumstances surrounding a sudden death may indicate that heroin had been ingested by the deceased. Subsequent analysis might dispute this. It would then be necessary to conduct additional testing to identify the cause of death. This would require additional discussions with the customer—in this case, the police and the courts.

### Principle 2: Analytical Measurements Should Be Made Using Methods and Equipment That Have Been Tested to Ensure They Are Fit for Purpose

If you want to use a method to analyse cocaine, then you need to have demonstrated and documented that it works and that it produces accurate and reliable results within a stated range. If your method uses the technique of HPLC, then you must be able to demonstrate that the instrument has been calibrated to a suitable level within the scope of your method. These processes are discussed in greater detail in the sections on *method validation* and *equipment qualification*.

### Principle 3: Staff Making the Measurements Should Be Qualified and Competent to Undertake the Task

When undertaking any forensic examination, the analyst should have an adequate understanding of the principles and methodologies associated with that examination. The analyst should feel comfortable enough to assess the risks associated with using the method and should be able to interpret the data outputs correctly. To ensure that this is carried out, a series of documented training programmes should be devised and implemented within the laboratory. All staff involved in the analysis must have received training prior to reporting any results. This is discussed further in Chapter 9.

## Principle 4: Regular Independent Assessment of the Technical Performance of a Laboratory Should Be Made

How would you measure up if another laboratory repeated your analysis? With any method, there are a number of potentially complex steps to undertake and, no matter how well trained the analyst is, mistakes do happen. One way to increase the level of confidence within the laboratory is to compare your results with those of other laboratories by participating in a proficiency testing scheme (PTS). External feedback from several sources, whether it is positive or negative, can help to improve your own processes. This is discussed further in Chapter 9.

## Principle 5: Analytical Measurements Made in One Location Should Be Consistent with Those Made Elsewhere

Imagine that you are working with one of the forensic science analytical testing companies and have just undertaken a piece of work at the request of the crown prosecution service. You have analysed a brown powder and found it to contain 70% of the controlled drug diamorphine (heroin). The counsel for the defence has requested that the sample be analysed by an independent laboratory of its choosing. The sample is reanalysed and there is a dispute in relation to the result because the defence analyst finds only 10% of the controlled drug diamorphine. You attend court and are asked to justify your results. Your lab has a rigorous quality assurance system that includes method validation, system suitability, and instrument qualification; the defence lab does not. Your result is accepted as the correct value.

## Principle 6: Organisations Making Analytical Measurements Should Have Well-Defined Quality Control and Quality Assurance Procedures

The preceding example demonstrates the importance of having well documented quality assurance procedures in place in the laboratory. Many laboratories choose to seek accreditation by way of a formal quality management system such as ISO/IEC 17025, which is available by application to UKAS (United Kingdom Accreditation Service). The processes involved and the choices available will be discussed in greater detail in Chapter 9.

These six principles provide a starting point for ensuring reliability within the laboratory and hence with the results generated. The aim of the information within this chapter is to help you understand the need for certain elements within a quality system, rather than to provide you with information to reproduce them. In order to be able do that and to find out more information about quality systems, individuals are directed to Chapter 9 and to the wealth of information available in quality standard documentation—for example, ISO/IEC 17025 or the Eurachem guides. Lots of information is available from many sources in relation to ensuring quality of data and further reading is documented at the end of the chapter.

## Estimation of Uncertainty of Measurement, Sources of Error, and Statistical Evaluation

It would be fair to say that an analysis can never be totally free from error or uncertainty in relation to the quoted result. Although the error cannot be eliminated, we can at least demonstrate to the courts that we understand the scale of the error associated with our results and have taken steps to control and measure it. The BS EN ISO/IEC 17025: 2005 standard document (*General Requirements for the Competence of Testing and Calibration Laboratories*) provides guidance and states that 'testing laboratories shall have and shall apply procedures for estimating uncertainty of measurement'. Eurachem/CITAC has produced a document (*Qualifying Uncertainty in Analytical Measurement*) that provides a comprehensive guide to measuring uncertainty and subsequent reporting of the data.

The uncertainty of measurement defines the range of values that might reasonably be expected from a measured quantity, whether that is flow rate or peak area within HPLC or whether it is a reported finding as a result of a forensic investigation. By measuring the uncertainty in a series of measurements, we can assign a level of confidence or reliability to our own results (i.e., how different our calculated value is from the true value). In terms of reporting forensic data, this could be represented as a percentage of controlled drug within an amount of powder.

For example, an analyst might examine a powder by HPLC and find that it contained 20% ± 1% of the controlled drug amphetamine. The uncertainty in this case has been determined to be 1%; therefore, the actual result might vary between 19 and 21%. Consider another example: If we look at a column thermostat on an HPLC system that has been set at 25°C, then we might expect that the temperature will remain at 25°C or close to this figure. To a large extent, this will be the case; however, there will be subtle temperature fluctuations with time. It is therefore necessary to assess the level of variation within the temperature readings in order to assess the uncertainty in the measurement as well as the impact on any results generated using this piece of equipment. A common way to conduct this type of measurement is to take multiple readings of the same variable over a specified period of time. If we record the temperature in our column thermostat over a 48-hour period and plot the results, we will be able to measure the level of error within the

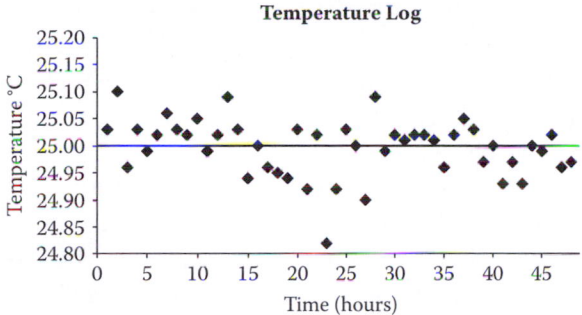

**Figure 8.1** Log of temperature values.

data set and assign a level of confidence around our temperature of 25°C (see Figure 8.1).

We can see from the results in Figure 8.2 that some of our measurements correspond to exactly 25°C and that some do not. The results range from 24.82 to 25.10°C and the average is 25.00°C. We are now able to measure the level of error in relation to our temperature by applying a level of probability. If we validate our method with an acceptable error range of 25°C ± 1.0°C and are happy with the results in terms of resolution, retention time, etc., we can be sure that temperature fluctuations due to this systematic thermostat error have been controlled effectively and the impact on the result is negligible.

Before carrying out any analysis, we must decide what level of error we are willing to accept and what level of error those judging our data will be

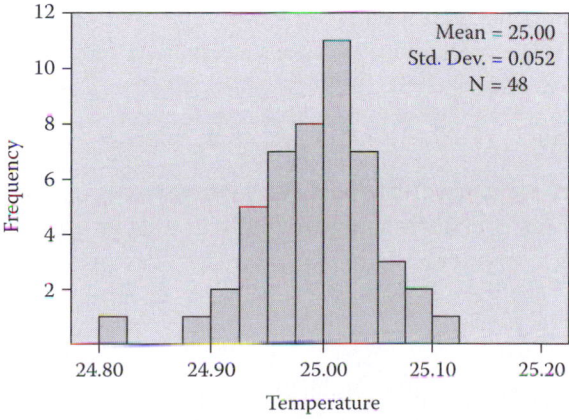

**Figure 8.2 (See colour insert following page 142.)** Normal distribution of temperature data set.

willing to accept. In forensic science, it might be acceptable to report a result for an examination of white powder for drugs to ±2% in terms of percentage of cocaine; however, it would not be acceptable to report that the powder 'might' contain cocaine. A small error in percentage recovered is acceptable because the Misuse of Drugs Act does not, in most cases, distinguish between such small variations in content. On the other hand, the act does make huge distinctions between controlled and noncontrolled drugs and within the different classification of controlled drugs. Therefore, an error in relation to what is present in the sample from a controlled drugs perspective is not acceptable.

This might help to explain why forensic analysts invest a lot of time in the identification or qualitative stage of the analysis. It is not uncommon to find forensic laboratories performing a three-tier assessment of potential controlled substances. This will involve a presumptive test, followed by a chromatographic separation and finally a mass spectral identification. Many of the analyses performed in the forensic environment are focused on what is present in the sample, rather than how much is there. Another example of this might be a paint examination, where it is important to assess the components present in relation to their chemical composition and also layer structure. Toxicological analysis, on the other hand, does require both a qualitative and quantitative assessment of what is present in the sample.

Uncertainty in measurement can arise from many different sources because all of the activities that go into generating the result are subject to some variation. The first type of variation is called *random error* and is related to the amount of scatter within a data set. The second type of variation is related to differences between the mean value for the data set and the accepted value for the measurement; this is referred to as *systematic error*. The total uncertainty of measurement for a particular measured quantity comprises the random element and the systematic element.

## Random Errors

Random errors are caused by uncontrollable variables within the measurement system and can be assessed through multiple measurements of the same sample and assessing the relative standard deviation.

## Systematic Errors

Systematic errors can usually be identified through careful investigation and therefore have an assignable cause and are constant for a particular data set determined under the same set of conditions. For example, a systematic error in detector wavelength will have a uniform impact on all measurements made under the same conditions in the same chromatographic sequence. Systematic errors can arise from the following broad sources:

- the instrument
- the method
- the analyst

Any instrument used in the lab will have a level of error associated with it. The term 'instrument' refers to any piece of equipment capable of making and/or recording a measurement. In HPLC, the whole system can be regarded as an instrument, but we must look at all of the individual components within the system to assess the total uncertainty of measurement. This would include the pump, the injector, the detector, and the data-recording device. Systematic errors due to the instrument are assessed as part of the installation and ongoing instrument qualification processes. Method errors can arise when there is ambiguity within the standard operating procedure (SOP). Systematic errors due to the method are assessed as part of the method validation and system suitability processes using certified reference materials obtained from authentic sources. Analyst errors can be the result of carelessness or inappropriate training of individual scientists within the laboratory.

In addition to this, systematic errors can also be introduced during the sample preparation. An example of this might be possible errors relating to use of volumetric glassware and volumetric pipettes that might arise as a result of the way in which an analyst uses volumetric glassware (i.e., the way in which the final volume is adjusted and read on a volumetric flask). Some analysts might read the final volume above the meniscus and some below. This would not be considered a significant source of error when the same analyst is preparing both samples and reference standards for an assay. It can become problematic however, when there is a requirement for different analysts to perform different sections within an analysis. This is often required to ensure that contamination issues are eliminated in forensic science. Systematic errors due to the analyst are assessed as part of the method validation and system suitability processes. Good practice and specialised training programmes that form part of a laboratory quality management system (see Chapter 9) help to minimise such errors.

Exactly how the information is used in relation to uncertainty measurement is dependent on the laboratory environment and the level of accreditation or quality management programmes that exist within it. Individual laboratories will have quality systems and standard operating procedures that deal with the measurement and recording of uncertainty measurement. To deal with all of the possible outcomes and eventualities is beyond the scope of this book. Suffice to say that it is important to be aware of the requirement and reasoning behind the application of uncertainty measurement.

## Instrument Qualification

The instrumentation comprises the individual elements that make up the HPLC system, including the pump, injector, column compartment, and detector. Qualification is performed on any element that, if defective, will have a significant effect on the result. Each of these elements must be assessed against a predetermined specification before it can be deemed fit for purpose and used to measure 'real' samples. The specifications used in the qualification of instrumentation are usually provided by the individual vendors of the equipment. Care should be taken because not all manufacturers' specifications are the same.

As you read on, you will see that choice of the vendor in itself forms part of the assessment process. Instrument qualification forms the basis of any quality programme in relation to the generation of data and there must be documented evidence that the systems are properly maintained and calibrated. For example, it is not good practice—not to mention good use of resources—to spend time validating a method using certified reference materials on an instrument that has never been qualified. This would render the results invalid at best and open to scrutiny in court.

Instrument qualification can be carried out in a number of ways, but the most widely documented appears to be the 4Q model. Credit for this model is given to Bedson and Sargent (1996) and a guide (*The Development and Application of Guidance on Equipment Qualification of Analytical Instruments*) is available through the National Measurement System Chemical and Biological Metrology Web site. The 4Q model is based on four stages of instrument qualification:

- Design qualification (DQ) is assessed before the purchase of an instrument and looks at different suppliers' products' specifications against laboratory requirements and ensures that the environment is suitable for instrument use.
- Installation qualification (IQ) ensures that all of the equipment has arrived per the request and is in good working order. It also verifies that the instrument is installed per the manufacturer's instructions.
- Operational qualification (OQ) provides evidence to support operational function and performance against a set of predefined specifications.
- Performance qualification (PQ) ensures that the system continues to perform at an acceptable level against predefined criteria and includes routine and preventative maintenance.

Instrument qualification is an ongoing process that is represented in Figure 8.3.

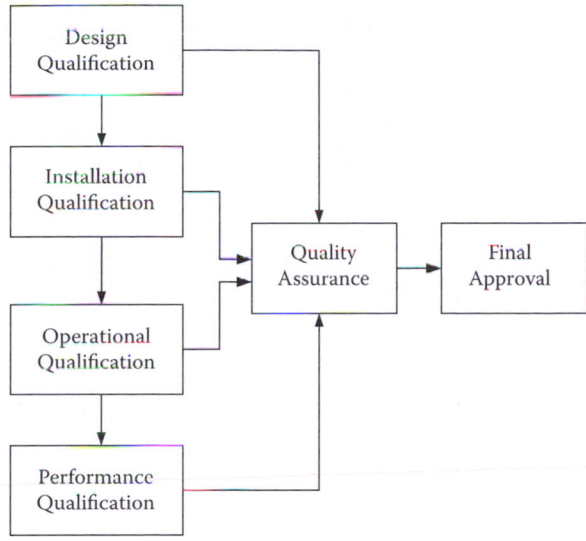

**Figure 8.3** Instrument qualification.

All of the qualification processes should be planned and documented for future reference. The exact nature of the documentation will be at the discretion of the individual laboratory, but the following should be considered as a minimum requirement:

1. Qualification plan.
2. DQ documentation containing information relating to the functions and performance of the system, an assessment of the environment, and an assessment of the vendor. During the DQ phase, the laboratory requirements are compared with the manufacturer's instrument specifications. At this stage, we are looking at the functionality of the instrument in the broadest of terms. An example of this might be whether the system includes a diode array detector or a simple fixed wavelength detector. Some of the questions that need to be addressed and documented at this stage are as follows:
   - What assays will the laboratory be performing? This is very important in a forensic environment due to the nonroutine, diverse nature of the testing.
   - Will this require specialised equipment? For example, will a low-flow pump, quaternary pumping system, large sample volume, column heater, etc. be needed?
   - How much can be spent?
   - Are all of the amenities available? For example, are there a suitable number of conveniently located electrical sockets?

- Is there enough bench space in the laboratory and can the instrument be accessed conveniently?
- Can waste solvents be easily collected and disposed of?
- How will the vendor be assessed in terms of reliability?
- Which vendor is easy to work with (including emergency repairs)?

Although this is not an exhaustive list, all of these questions must be answered satisfactorily before an instrument and vendor are chosen. All information needs to be documented to demonstrate that the vendor and the instrument were chosen with due care and attention to laboratory needs.

3. IQ documentation verifying that the correct instrument has been delivered and that it is in good working order. The installation documentation should record that the system has been installed per the manufacturer's instructions and in the correct environment. Some of the information that should be obtained and documented includes the following:

- Prepare the environment within the laboratory per the manufacturer's specification. An example might be to ensure that space is sufficient to allow the free flow of air around the instrument to avoid overheating or to ensure that there is sufficient room on the bench to place the instrument to create a comfortable work space.
- Check that the correct equipment has been received against the original signed order. Go through all of the parts that have been received and check them off against the order. Retain this information.
- Check the instrument for obvious damage.
- Ensure that all of the documentation (manuals, maintenance instructions) has been supplied. Some of this might be in disk format.
- Ensure that all of the components switch on when power is supplied.

4. OQ documentation verifies that the instrument performs to the manufacturer's specification and to any in-house specifications. Some of the parameters that might be included in a typical OQ for an HPLC system would include the following tests:

   a.  flow rate
   b.  gradient
   c.  proportioning valve
   d.  injector—volume accuracy, repeatability
   e.  detector wavelength accuracy
   f.  holmium filter test
   g.  lamp intensity profile
   h.  baseline noise

5. PQ testing demonstrates that the system is consistently performing to the manufacturer's specification and that it is appropriate for its intended use. PQ testing is performed on an annual basis or more frequently if deemed appropriate by the user. If the testing is performed on an annual basis, all of the OQ testing should be performed on each occasion. More frequent testing should also be performed depending on the nature of the instrument and the acceptance criteria within the specification. Time intervals for additional testing should be chosen such that there will be a high probability that the testing will meet with the specification. If this does not happen and the acceptance criteria are not met, then all of the data generated using that piece of equipment from the date of the last PQ test will be questionable and will require a thorough investigation.

This is not to say that limits should be set so wide that passing is assured every time. The specification limits should challenge the instrument but should also be realistic. Instrument manufacturers can give guidance where required. PQ testing should also be performed when instrumentation has been moved, there has been a repair to the system, or preventative maintenance procedures have been carried out (see Chapter 10). For example, if the pump seals have been leaking and a repair has been carried out, then the pump should be tested to ensure flow rate accuracy. Instruction in relation to PQ testing following repairs should be documented clearly in the instrument standard operating procedure. All of the information relating to the PQ should be documented and filed away in a secure manner for future reference. The exact nature and time line for the storage will be governed by the company quality system.

## Method Validation

### Introduction and Validation Procedures

Method validation is a way of ensuring that the analytical method used in a specific test procedure is suitable for that task or fit for purpose. It is required in a variety of different laboratory environments where there is a public interest in the outcome of the analyses (e.g., pharmaceutical laboratories, cosmetics laboratories, forensic laboratories, food safety laboratories). According to Peters, Drummer, and Musshoff (2007), validation is particularly important in a forensic environment—not only to ensure that data are reliable but also to ensure that there are no 'unjustifiable legal consequences' for the defendant in court. Method validation within the forensic arena is usually applied to two types of analytical procedures:

- *Quantitative test for the active substance.* This describes a situation where there is a requirement to assess how much of a substance is present in a mixture. It may be necessary to quantify several different substances within the same sample mixture. An example of this would be an unknown powder thought to contain both heroin and codeine.
- *Identification (qualitative) tests.* Identification testing is used to identify which compounds are within the sample matrix. This is usually carried out by comparison to a reference standard.

Validation involves applying a set of predetermined parameters and limits to methods of analysis and assessing the data to ensure suitability for use. Validation should always be conducted prior to using the method in a commercial or regulated environment. The validation limits must be met in order for the method to be considered suitable for use. The outcomes from the validation give an indication of the method's quality, reliability, and consistency of use.

Well characterised reference standard materials representative of all of the expected components must be used in all phases of the validation procedure. ISO (International Organisation on Standardisation) guide 30:1992—*Reference Materials Part 1. Guide to Terms and Definitions Used in Connection with Reference Materials*—defines a reference material as being a 'material or substance one or more of whose property values are sufficiently homogeneous and well established to be used for the calibration of an apparatus, the assessment of a measurement method or for assigning values to materials'. Reference materials can be purchased from a number of sources and are accompanied by a certificate of analysis or a reference material certificate, which must contain all of the information essential to its use.

A reference material certificate is defined by ISO 31:2000—*Guide to the Contents of Certificates and Label of Reference Materials*—as a 'document accompanying a certified reference material stating one or more property values and their uncertainties, and confirming that the necessary procedures have been carried out to ensure their validity and traceability'. For example, this would include information relating to the batch number, a description of the material, the concentration of the material, storage conditions, and expiry date. The content of the certificate is discussed in detail in ISO 31:2000—*Guide to the Contents of Certificates and Label of Reference Materials*.

The properties of reference standards can vary and care should be taken to ensure that the standard chosen meets the requirements of the testing. It would not be appropriate, for example, to use a reference standard that claims to be >95% pure in an assay used to determine the concentration of controlled drug in a suspect sample. It would be more appropriate to choose a reference standard that reports an exact concentration, including the measurement of uncertainty, or, at the very least, that claims to be >99% pure. Using the first reference standard would introduce a substantial level of error into the measurement. Further information relating to reference materials can be found on the British Standards Web site in relevant ISO documents and guides and in the Eurachem guide EEE/RM/062rev3 2002: *The Selection and Use of Reference Materials—A Basic Guide for Laboratories and Accreditation Bodies*.

Method validation is required in a number of scenarios and should be considered as a requirement under the following conditions:

- before the introduction of a new analytical method for routine use
- when the conditions used to perform an existing validated method have changed (e.g., a change in temperature or extraction procedure)
- when the scope of the method has changed (e.g., the method is being applied to a different product or sample matrix)

Before the method validation exercise begins, a number of different things need to be considered. First, it is assumed that the method has been developed and that it meets the intended scope (see Chapter 6). Collaboration will be required with other departments to ensure that this has been carried out successfully. Second, the validation protocol must be prepared and acceptance criteria agreed upon with those responsible for the method development (e.g., whether it is critical that the level of intermediate precision be 2%). It is always a good idea to run a number of preliminary experiments before embarking on the validation proper in order to ensure that the method is behaving as intended. Some minor adjustments can be made at this stage that can save time and effort in the long run. Third, a standard operating procedure should be prepared and should be followed during the validation process.

## Validation Protocol

The validation protocol sets out the testing requirements and the criteria that need to be met prior to the method being considered suitable for use. The scope of the method should have been defined during the development phase and the criteria used to assess the method will be based around results generated during the early development phase. Before the protocol is written, the following questions need to be answered.

### What Concentration Levels Are Expected for the Analyte?

In a forensic environment, it is often not easy to answer these questions due to the uncertain and wide ranging natures of samples received within laboratories. Heroin samples, for example, range in purity; therefore, it would be difficult to validate a method with an acceptable range that would cover all eventualities. This can be overcome by using information 'intelligence' and validating a method around the mean of the values obtained. Samples that fall outside the top end of the validated range can be diluted so that they fall within the range of the method. Samples that fall outside the lower end of the method range can be concentrated or prepared so that they fall within the range.

### Will There Be Any Interfering Substances?

An interfering substance may be defined as any other unknown substance that might be present in the sample matrix that could, in theory, interfere with the detection and quantitation of the analyte of interest. Within the forensic arena, this might include substances like sugars, paracetamol, or topical anaesthetics (e.g., lignocaine). Gathered 'intelligence' from previous analysis can help to answer questions in relation to the interfering substances that might be expected at a particular time. It is worth noting that, due to the diverse nature of cutting agents, validations may have to be readdressed on a regular basis. The method should be validated using common cutting agents to simulate the matrix.

### Is the Method Intended for Qualitative and/or Quantitative Analysis?

The extent of validation required will depend on the nature of the intended analytical outcomes. Is the method quantitative (i.e., to allow a measure of quantity of some characteristic of the sample, whether it be active ingredient, impurities, or excipient) or is the method simply a determination of what is present (qualitative)? In the forensic environment, both qualitative and quantitative data will most likely be required; therefore, it is advisable to ensure that the validation covers both aspects. The extent to which validation is conducted and the parameters that should be included have been the subject of several guidance documents produced by a variety of industrial committees and regulatory or advisory bodies.

The International Conference on Harmonisation of Technical Requirements for Registration of Pharmaceuticals for Human Use (ICH) has published a number of guidelines relating to validation of methods aimed at the pharmaceutical and biopharmaceutical sector. Its purpose is to bring together regulatory authorities from Europe, Japan, and the United States and industry experts to discuss scientific aspects of product registration and to disseminate best practice. Although the documents are sector focused, the themes and the procedures are relevant to the forensic industry as well (2005 ICH Harmonised Tripartite Guideline Q2(R1): *Validation of Analytical Procedures: Text and Methodology*).

**Table 8.1   Validation Requirements Guideline**

|  | Controlled Drug—Identification (Qualitative) | Cutting Agents—Impurities (Quantitative) | Cutting Agents—Impurities (Qualitative) | Controlled Drug—Assay (Quantitative) |
|---|---|---|---|---|
| Accuracy | x | √ | x | √ |
| Precision | x | √ | x | √ |
| Linearity | x | √ | x | √ |
| Specificity | √ | √ | √ | √ |
| Range | x | √ | x | √ |
| LOD | x | x | √ | x |
| LOQ | x | √ | x | x |
| Robustness | x | √ | x | √ |

The United States Food and Drug Administration (U.S. FDA) has published a number of guidance documents (May 2001, *Guidance for Industry: Bioanalytical Method Validation,* and August 2000, *Guidance for Industry: Analytical Procedures and Methods Validation, Chemistry, Manufacturing and Controls Documentation,* Draft Guidance) aimed at the pharmaceutical and biopharmaceutical industries. The latter of the two referenced documents is very comprehensive in its descriptions and gives a lot of good information in relation to which types of techniques should be used to achieve certain outcomes and which aspects of the techniques need to be addressed during method development and validation. The FDA's responsibilities lie primarily in ensuring public health in the United States through safe and effective medicinal products by monitoring processes during the product development phase as well as beyond product release. Much of the material covered mirrors other referenced documentation and can be easily adapted for use within the forensic arena.

ISO/IEC 17025 includes a section (2005: Section 5.4 *Test and Calibration Methods and Methods Validation*) dedicated to the test and calibration of methods including method validation. Much of the material already covered by the ICH and FDA is again discussed, as is uncertainty in measurement, which will be discussed in this chapter in more detail later. Emphasis is placed on continuous review of validation data to verify that the intended purpose of the method is still being achieved. Table 8.1 can be used as a guideline for the required validation elements.

## What Level of Precision and Accuracy Is Required?
Levels of precision and accuracy should be determined in conjunction with other testing laboratories providing the same services and with the legal systems operating in the region. Pharmaceutical validation offers the highest level of accuracy and precision. Therefore, setting this as a benchmark from which to work would be considered reasonable.

## How Robust Must the Method Be?

This will depend on a number of factors, which might include the following:

- analyst competency
- whether the method has to be run in different labs
- whether the method has to be run using different instruments

Once all of the preceding issues have been addressed, the protocol should be prepared and approved by laboratory management. It is management's responsibility to ensure that the protocols meet the needs of the laboratory and the appropriate regulatory body (if applicable) and that the resulting method is fit for purpose (quality systems will be discussed in more detail in Chapter 9). The parameters mentioned in Table 8.1 are discussed in more detail later with some examples of the limits that might be applied.

## Validation Elements

### Accuracy

According to the ICH Guidelines (Q2 reference), accuracy can be described as 'the closeness of agreement between the value which is accepted as a conventional true value or an accepted reference value and the value found' (Figure 8.4). The accuracy of the method should be established by preparing test mixtures to which known amounts of the drug material have been added. In forensic science laboratories, it may not be possible to establish all of the possible components within a particular sample; therefore, it may be sufficient to establish the accuracy of the method using only pure drug substances. Specificity must then be established using a variety of common cutting agents.

Accuracy of the method should be established using at least nine determinations of the drug in question over a minimum of three concentrations covering the specified range of the method. This would normally be represented as follows:

three replicates at 70% target concentration
three replicates at 100% target concentration
three replicates at 130% target concentration

Each of the replicates should be prepared independently and the accuracy calculated and reported as a percent of recovery of the target value using reference standards. Determination of the target value can be difficult in a forensic environment; however, gathering intelligence from other providers or via network groups can be helpful. Limits would depend on the nature of the testing and the levels that are representative of 100%.

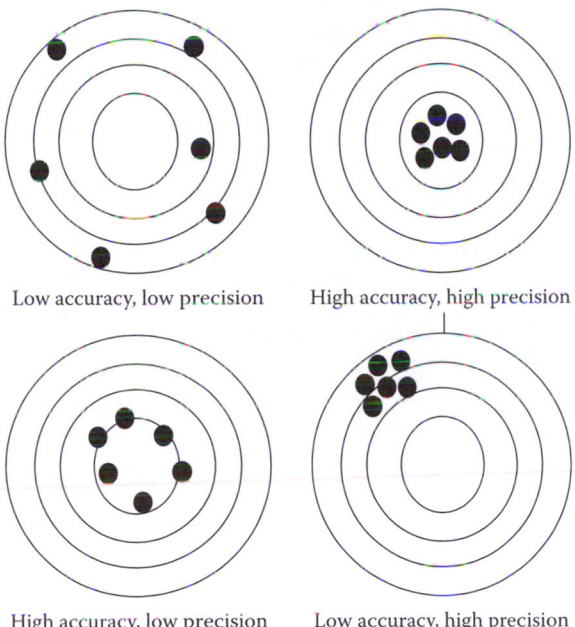

Low accuracy, low precision    High accuracy, high precision

High accuracy, low precision    Low accuracy, high precision

**Figure 8.4** Measurement of accuracy and precision.

### HYPOTHETICAL EXAMPLE 8.1

You are asked to determine the accuracy of a method for the determination of amphetamine using HPLC. Based on 'intelligence' gathered from local and national drug squads and from your own analytical experience, you expect to find street samples containing anywhere from 10 to 30% w/w amphetamine. You know from your method development analysts that a solution of amphetamine containing 100 µg/mL gives an acceptable peak area using HPLC (see Chapter 10). You assign your 100% target value at 100 µg/mL. In order to complete the accuracy testing, you would prepare three independent solutions using verified reference standard at 70, 100, and 130% of the target value and run these through the HPLC system in duplicate, recording the peak area. Table 8.2 provides an example of one way in which accuracy data can be elucidated and recorded for a particular method.

## Precision

Precision is expressed as the closeness of agreement between a set of replicate results obtained from multiple analyses of the same homogeneous solution. There are three levels of precision measurement:

- repeatability
- intermediate precision
- reproducibility

*Repeatability* measures the precision of the method under the same set of conditions over a short period of time. Precision measurement is usually

**Table 8.2  Example of Accuracy Results**

| Sample | Concentration | Injection No. | Area | Mean | Recovery (%) | Mean Recovery (%) | SD | RSD (%) |
|--------|---------------|---------------|------|------|--------------|-------------------|-----|---------|
| 70% | 70 µg/mL | 1 | 350 / 355 | 353 | 100.22 | 70.26 | 0.26 | 0.37 |
| | | 2 | 351 / 353 | 352 | 100.09 | | | |
| | | 3 | 354 / 355 | 355 | 100.80 | | | |
| 100% | 100 µg/mL | 4 | 502 / 501 | 502 | 99.83 | 100.03 | 0.17 | 0.17 |
| | | 5 | 503 / 503 | 503 | 100.13 | | | |
| | | 6 | 501 / 505 | 503 | 100.13 | | | |
| 130% | 130 µg/mL | 7 | 654 / 655 | 655 | 100.22 | 130.16 | 0.51 | 0.39 |
| | | 8 | 655 / 657 | 656 | 100.45 | | | |
| | | 9 | 650 / 652 | 651 | 99.68 | | | |

expressed as a relative standard deviation (RSD) obtained from at least one set of between 5 and 10 measurements of the same solution and/or from the determination of accuracy (see Tables 8.3 and 8.4).

**HYPOTHETICAL EXAMPLE 8.2**

You have been asked to determine the repeatability for a particular assay at the 100% target value. Prepare a standard solution containing the correct amount of analyte and run the solution six times in succession through the HPLC system, recording the values for peak area. Tabulate your results and calculate the RSD. Data from the accuracy section of the validation is also used to assess the precision of the assay by examining the percent of RSD obtained from all of the results.

*Intermediate precision* measures the precision of the method when conducted within the same laboratory and when some of the physical conditions may change (e.g., a different analyst, a different day, or a different instrument).

Table 8.3   Repeatability Derived from Six Injections of the Same Solution

| Injection No. | Peak Area | SD | RSD (% CV; $n = 6$) |
|---|---|---|---|
| 1 | 500 | | |
| 2 | 503 | | |
| 3 | 506 | 1.72 | 0.43 |
| 4 | 501 | | |
| 5 | 503 | | |
| 6 | 505 | | |

Note: Limit (% RSD) ≤ 0.5.

Table 8.4   Repeatability Derived from the Accuracy Data

| Sample | Concentration | Recovery (%) | Mean Recovery (%) | SD | RSD (% CV; $n = 9$) |
|---|---|---|---|---|---|
| 70% | 570 µg/mL | 100.22 | | | |
| | | 100.09 | | | |
| | | 100.80 | | | |
| 100% | 100 µg/mL | 99.83 | 100.17 | 0.33 | 0.32 |
| | | 100.13 | | | |
| | | 100.13 | | | |
| 130% | 130 µg/mL | 100.22 | | | |
| | | 100.45 | | | |
| | | 99.68 | | | |

Note: Limit (% RSD) ≤ 0.5.

Precision measurement is expressed in terms of the percent of RSD of the data obtained from the accuracy section of the validation performed by different analysts on different days using different equipment. The determination of accuracy (three concentrations) would be performed over several days with different operators using the same set of homogeneous standard solutions (see Table 8.5). This can be very time consuming; therefore, a balance needs to be struck among the number of measurements, analyst time, and cost to the business. If only one type of instrument will be used, then this part can be eliminated.

*Reproducibility* measures the precision of the method between different laboratories. This would only be required when there is a genuine need to have the assay performed at more than one location. The precision measurement

**Table 8.5  Intermediate Precision at the 100% Concentration**

| Analyst | Day | Mean Recovery (%) | SD | % RSD (% CV; $n = 27$) |
|---|---|---|---|---|
| 1 | 1 | 100.02 | | |
| | 2 | 100.1 | 0.040 | 0.04 |
| | 3 | 100.05 | | |
| 2 | 1 | 100.12 | | |
| | 2 | 100.08 | 0.067 | 0.07 |
| | 3 | 99.99 | | |

*Note:* Limit (% RSD) ≤ 0.5.

would be generated by assessing the RSD from accuracy data independently generated by both laboratories. Again, the same set of homogeneous standard solutions would be used by both laboratories.

## *Linearity*

The linearity of the method is a measure of the procedure's 'ability (within a given range) to obtain test results which are directly proportional to the concentration (amount) of the analyte in the sample' (ICH Guideline Q2(R1)). In order for the assay to be considered both accurate and precise, the response from the detector must be directly proportional to the concentration within a given concentration range.

Linearity is measured by calculating the regression line of a plot of detector response against concentration using the equation $y = mx + c$ (see Figure 8.5). The linearity of the method should be demonstrated over at least three concentrations (accuracy data), but ideally should be measured over five concentrations, with additional measurements made at the lower end of the range to accommodate samples that fall short of the 100% value.

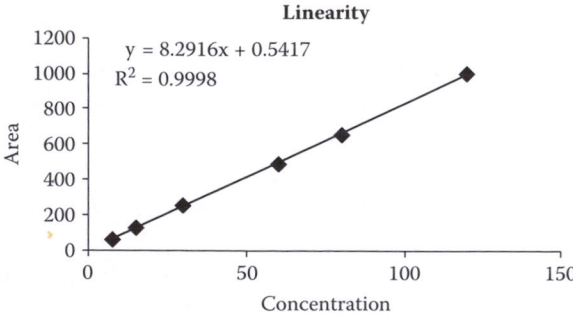

**Figure 8.5** Linear plot showing regression line and correlation coefficient.

Data submitted should include the slope, intercept, and correlation coefficient information.

A value of ≥0.999 would be considered acceptable for the correlation coefficient. It should be noted, however, that the plot must be examined visually for any slight deviations from the line at low and high levels (see Chapter 10). The y-intercept should be assessed and should be less than, at maximum, 4% of the response obtained with the 100% analyte concentration.

### HYPOTHETICAL EXAMPLE 8.3

You have been asked to determine the linearity for a method designed to measure the amount of a controlled drug within drug wraps. You have determined a 100% value for your assay of 100 µg/mL. You prepare five solutions containing concentrations of material at 10, 20, 50, 100, and 130% of the target concentration, respectively. These five solutions should be prepared in triplicate and the accuracy at each concentration should be determined using the previously mentioned technique.

## Range

The range of an analytical method is the concentration interval over which the method is acceptable in terms of accuracy, linearity, and precision.

### HYPOTHETICAL EXAMPLE 8.4

In the preceding examples, the range of the method would be 10–130% of target analyte concentration (100 µg/mL) or 10–130 µg/mL.

## Limit of Detection (LOD)

The limit of detection (see Figure 8.6) of an analytical method is the lowest concentration of the analyte that produces a response above the noise level of the HPLC system. An acceptable noise level is considered to be typically three times the noise level. In practical terms, the limit of detection can be calculated by measuring the signal-to-noise ratio of a series of sample dilutions until such time as the signal-to-noise ratio exceeds 3:

$$LOD = C_s \frac{3}{S/N}$$

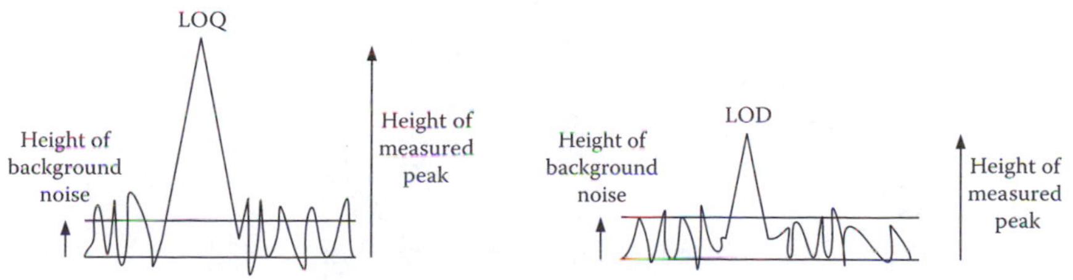

**Figure 8.6** LOQ and LOD measurements.

where

$C_s$ = concentration of the sample
$S/N = h_s/2h_n$
$h_s$ = height of measured peak
$h_n$ = height of background noise

**HYPOTHETICAL EXAMPLE 8.5**

You have been asked to establish the limit of detection for a particular compound in your lab. You would prepare a series of sample dilutions at concentrations in the range of 0.5–0.05% of the target concentration. These solutions would be sequentially injected until such time as the signal-to-noise ratio exceeded 3. The concentration value immediately above this value would be considered to be the limit of detection.

## Limit of Quantitation (LOQ)

The limit of quantitation (see Figure 8.6) of an analytical method is the lowest concentration of the analyte that can be accurately and precisely measured. The limit of quantitation can be determined in one of two ways:

- A series of concentrations (usually in the region of 0.01–0.5% of the target concentration) of the analyte can be prepared (in triplicate) and the accuracy of each solution measured. The LOQ is reached when the data no longer fall within the predefined limits for accuracy. The limits set for accuracy around the LOQ would be in the region of approximately ±20%, depending on the nature of the sample.
- The signal-to-noise ratio of the system can be measured. A signal-to-noise ratio of 10:1 is considered acceptable. A similar experiment would be conducted as that performed in the LOD measurement until such time as the signal-to-noise ratio exceeds 10%. The value immediately above this value would be considered to be the limit of quantitation:

$$LOD = C_s \frac{10}{S/N}$$

where

$C_s$ = concentration of the sample
$S/N = h_s/2h_n$
$h_s$ = height of measured peak
$h_n$ = height of background noise

## Robustness

The robustness of a method is a measure of the method's ability to remain unaffected by small but deliberate changes to the method parameters (e.g., flow rate, detector wavelength, etc.; see Table 8.6 for some further examples).

**Table 8.6   Robustness Testing**

| Parameter | Target | Robustness Testing |
|---|---|---|
| Flow rate | 1 mL/min | 0.9, 1.0, and 1.1 mL/min |
| Column temperature | 40°C | 35, 40, and 45°C |
| Mobile phase pH | 5.4 | 5.2, 5.4, and 5.6 |
| Mobile phase composition | 50% methanol/50% water | 48/52, 50/50, and 52/48 |
| Buffer concentration | 100 m$M$ | 90, 100, and 110 m$M$ |
| Injection volume | 10 μL | 9.9, 10, and 10.1 μL |
| Detector wavelength | 254 nm | 251, 254, and 257 nm |

The method must be evaluated against predefined acceptance criteria one parameter at a time or simultaneously as part of a design of experiment (DoE) approach (http://www.sigmaplus.fr). By identifying the parameters that are critical to the performance of the method, a range of values can be incorporated in the final version of the standard operating procedure or method:

*Flow rate.* The flow rate accuracy will be determined by the instrument used and will be documented in the supplier's operational qualification documentation. The impact of analyst error made during system setup should be validated and it would be appropriate to investigate the effect of small changes in the region of ±10% of the target flow rate. The flow rate may be adjusted by as much as 50%, provided that there are no adverse effects on the chromatography (i.e., resolution, peak shape and retention time). Common causes of flow rate error will be discussed further in Chapter 10.

*Column temperature.* This may be adjusted by as much as 20%, provided that there are no adverse effects on the chromatography (i.e., resolution, peak shape, and retention time).

*Mobile phase pH.* This may be adjusted by as much as ±0.2 pH units, provided that there are no adverse effects on the chromatography (i.e., resolution, peak shape, and retention time).

*Mobile phase composition.* This may be adjusted by as much as 30%, provided that there are no adverse effects on the chromatography (i.e., resolution, peak shape, and retention time).

*Buffer concentration.* This may be adjusted by as much as 10%, provided that there are no adverse effects on the chromatography (i.e., resolution, peak shape, and retention time). The pH of the solution, however, must remain within the designated effective range to ensure buffering capacity (see Chapter 3).

*Injection volume.* The injection volume variation will be determined by the instrument used and will be documented in the supplier's operational qualification documentation. The impact of analyst error made during system setup should be validated and it would be appropriate to investigate the effect of small changes in the region of ±10% of the target volume. Common causes of injector error will be discussed further in Chapter 10.

*Detector wavelength.* Detector wavelength accuracy will be determined by the instrument used and will be documented in the supplier's operational qualification documentation. The impact of analyst error made during system setup should be validated and it would be appropriate to investigate the effect of small changes in the region of ±3 nm from the target wavelength.

The limits for robustness testing will be established by assessing the effects of the small changes on the chromatography. This is established by measuring parameters such as resolution, peak tailing, and column efficiency and assessing these values against predetermined limits.

### HYPOTHETICAL EXAMPLE 8.6

You have been given a method for the quantitation of amphetamine in powders; this is able to separate amphetamine from other related amphetamine compounds such as methylamphetamine and common ecstasy type compounds such as MDA or MDMA. The resolution between your compounds should be no less than 1.2. The retention time is also used to identify any compounds of interest. Under normal method conditions, the resolution between MDA and MDMA is 1.4. You have been asked to assess the effects of a small change to the flow rate in relation to compound resolution and retention time in your assay. You would prepare a solution at 100% of the target concentration and run the repeatability section of the method validation at flow rates of 0.9 and 1.1 mL/min. You would measure the resolution in each case and assess this against the predefined acceptance criteria (not less than 1.2).

## Method Verification

Method verification could be considered to be a subset of the validation protocol. It is required when there has been a subtle but significant change to the method. This might include a situation where the concentration of a particular drug may have changed in terms of street levels and, as a consequence, the target concentration must change along with the method range. This would mean that elements of the validation protocol would need to be reassessed in light of the new situation. In this situation, the elements that would need reevaluation or verification would include accuracy, precision, and range.

Another situation where verification might be required would be when one lab developed and validated the method and then wanted another lab to carry out testing on its behalf. Although this can be referred to as method verification, it is usually better known under the banner of method transfer.

Compendial methods must also be verified for the particular application to which they are being applied. For example, a compendial method for the analysis of codeine in pharmaceutical-grade tablets will need to be verified in order to check elements such as specificity and range if it is to be used in a forensic application.

---

**KEY POINT SUMMARY**

**The Six Principles of Valid Analytical Measurement**

Principle 1: Analytical measurement should be made to satisfy an agreed upon requirement.

Principle 2: Analytical measurement should be made using methods and equipment that have been tested to ensure they are fit for purpose.

Principle 3: Staff making the measurements should be qualified and competent to undertake the task.

Principle 4: Regular independent assessment of the technical performance of a laboratory should be made.

Principle 5: Analytical measurement made in one location should be consistent with those made elsewhere.

Principle 6: Organisations making analytical measurements should have well defined quality control and quality assurance procedures.

**Uncertainty of Measurement and Errors**

Errors cannot be totally eliminated and there will always be some level of uncertainty around measurements. By applying the principles of VAM, we can ensure that the errors and uncertainty are kept to a minimum and accounted for in our final results.

**Instrument Qualification**

Performing instrument qualification satisfies VAM principles 2, 4, and 6.

Design qualification (DQ): is the instrument fit for purpose?

Installation qualification (IQ): have I installed the equipment correctly and in keeping with the manufacturer's guideline?

Operational qualification (OQ): does the system function as expected over a short period of time?

Performance qualification (PQ): does the system function as expected over an extended period of time?

*(continued on next page)*

---

**Method Validation**

A method validation strategy should form part of an overarching quality management system.

Performing method validation satisfies VAM principles 1, 2, 4, 5, and 6.

Accuracy: closeness of the measured value to the true value.

Precision: a measure of the variability of data.

Linearity: detector response is directly proportional to the sample concentration.

Range: the upper and lower values for concentration over which the method is linear, accurate, and precise.

Limit of detection (LOD): the lowest concentration of the analyte that produces a response above the level of noise.

Limit of quantification (LOQ): the lowest concentration of the analyte that can be accurately and precisely measured.

Robustness: measure of the impact from small but deliberate changes to the method parameters.

**Method Verification**

Method verification represents a subset of the validation parameters and should be undertaken when:

- a significant change has been made to the method
- a second laboratory wishes to perform the same testing using the same method (method transfer)
- there is a change to the sample matrix

QUESTIONS

1. What are the elements of validation required for a quantitative assay?
2. Why is it necessary to qualify analytical instrumentation?
3. List the 4Qs and describe the purpose of each.
4. What is meant by systematic error? Give two examples.
5. Give an example of a strategy that might be used to measure the accuracy of a method.

# References

Bedson, P., and M. Sargent. 2006. The development and application of guidance on equipment qualification of analytical instruments. *Accred. Qual. Assur.* 1: 265–274.

Peters, F. T., O. H. Drummer, and F. Musshoff. 2007. Validation of new methods. *Forensic Science International* 165: 216–224.

# Further Reading

## Books/Online Publications

Agilent Technologies. 2009. *Analytical instrument qualification and system validation* (www.home.agilent.com).

Barwick, V. J. 2003. *Preparation of calibration curves: A guide to best practice.* LGC.

British Standard, BS EN ISO/IEC 17025:2005. *General requirements for the competence of testing and calibration laboratories.*

Chan, C. C., Y. C. Lee, H. Lam, and X.-M. Zhang, eds. 2004. *Analytical method validation and instrument performance verification.* New York: John Wiley & Sons.

Eurachem/CITAC Guide CG-4. 2000. *Quantifying uncertainty in analytical measurement.*

Eurachem Guide EEE/RM/062rev3. 2002. *The selection and use of reference materials.*

European Agency for the Evaluation of Medicinal Products. 1995. *Validation of analytical procedures: Text and methodology.* ICH topic Q2 (R1) (available www.fda.gov; also available through CDER).

European Agency for the Evaluation of Medicinal Products. 1996. *Validation of analytical procedures: Methodology.* ICH topic Q2 B (available www.fda.gov; also available through CDER).

Hardcastle, W. A. 1998. *Qualitative analysis: A guide to best practice. Forensic science extension.* LGC.

ISO Guide 30. 1992. Reference materials part 1. *Guide to terms and definitions used in connection with reference standards.* PD 6532.

ISO Guide 31. 2000. *Guide to the contents of certificates and labels of reference materials.* PD 6532-2.

King, B. *In-house method validation: A guide for chemical laboratories.* LGC.

King, L. A. 2003. *Misuse of Drugs Act—A guide for forensic scientists.* Cambridge, England: Royal Society of Chemistry.

Miller, J. M., and J. B. Crowther, eds. 2000. *Analytical chemistry in a GMP environment: A practical guide.* New York: John Wiley & Sons.

Standard Committee for Quality and Competence, European Network of Forensic Science Institutes. 2006. *Validation and implementation of (new) methods QCC-VAL-001.*

*VAM principles.* National Measurement System Chemical and Biological Metrology Web site (www.nmschembio.org.uk).

## Journal Articles

Yazdi, A. S., and Z. Es'haghi. 2005. Surfactant enhanced liquid-phase microextraction of basic drugs of abuse in hair combined with high performance liquid chromatography. *Journal of Chromatography A* 1094: 1–8.

# Quality Systems

9

## Introduction

Quality or, more specifically, quality systems are important in environments where work is being carried out that is under strict regulation by external bodies such as the medicines and health regulatory authority (MHRA) regulating a pharmaceutical company. This is also true in forensic science laboratories. Although there is not a regulating body per se, it is the court or judicial system that requires the process be regulated. It is not mandatory in forensic science laboratories, but most, if not all, choose to introduce and maintain a quality system.

## What Is Quality?

Quality is a way of ensuring that the systems in place are 'fit for purpose' and making sure we get it 'right the first time'. This means that we should be able to ensure customer satisfaction, individual accountability, and integrity in the work; we should have a preset standard of work that is maintained and a way of ensuring continual improvement. Let us consider a generic forensic science process (Figure 9.1).

If we study this process to determine where issues of quality may be present, we find the following:

- *Scene examination.* If there is no standard procedure for the examination of a scene, it is possible that not enough of the space is examined or that essential pieces of evidence may go undetected due to the inadequate search of the area. It is also possible that a scene could be contaminated if it is not regulated to ensure that staff wear appropriate clothing and personal protective equipment and that access to the scene is restricted to certain individuals whose arrival and departure are logged. This can be avoided if general procedures are in place to ensure that these as well as other processes are controlled.
- *Recovery of evidence.* Consider a scenario in which a piece of evidence could prove vital to the case if found at the scene of a murder. The scene of crime officer is not fully trained and fails to recognise its

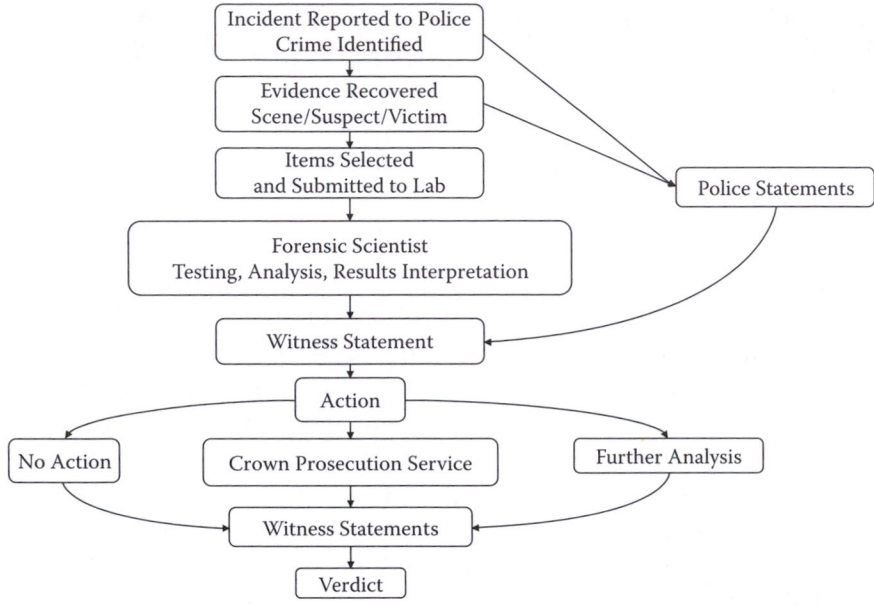

**Figure 9.1** General forensic process flow diagram.

potential value. If the object is incorrectly handled, the vital evidence that might be present could be lost. Again, we need adequate staff training and general procedures for the control of processes used by the crime scene officer.

- *Packaging of evidence.* Again, let us continue with the scenario that a murder has taken place. A hammer has been found near the victim; it has staining that has the appearance of blood and what looks like fingerprints in the red substance (assumed to be blood) on the handle. If the object is inappropriately handled and packaged at the scene, this potential evidence (the red staining and the fingerprints) could be lost due to interaction with the packaging or by not being packaged well enough to survive transport to the laboratory. Here we need to ensure that adequate staff training has occurred to equip the individual with the knowledge to make the decision for choosing appropriate packaging to ensure evidence preservation.
- *Continuity of evidence.* At the scene of the crime, the scene would be examined and evidence collected and packaged. Submissions are then transported to the laboratory for analysis; items will be examined and tested and results reported in the form of a statement that will be submitted to the Crown Prosecution Service (CPS), Procurator Fiscal (PF), or other intermediate organisation for processing for court. Consider one item—say, the hammer mentioned earlier—and

think about what happens at each step in the process. We must consider that, as forensic scientists, we will be expected to attend court and, if required, will give evidence in relation to the hammer that we have examined. If we cannot account for the whereabouts of the hammer at every step in the process of the examination—or, indeed, in the whole process from collection to submission to the court as evidence—we will look foolish (as will the process through which the item was examined) and subsequently may damage the whole investigation. This may result in the case being thrown out of court if the piece of evidence is a vital in the case.

We must ensure that we can account for every item in a case at all points in the investigation. This means that when we take receipt of an item from the submissions office, it must be signed and dated to say that it was intact at the time that we received it. When we commence our analysis of the item, we must complete the label on the packaging to show that we have accessed the item; during our analyses we will complete laboratory notes that will show what we did to the item, when and where the analyses took place, and when they were finished. When we have finished our work, we will return the item to its original packaging and signature-seal the item to show that it was sealed when we were finished. Then, we will return it to the submissions office where we will log the item. Should another scientist require access to the item, he or she will start the whole process over again.

- *Submission of evidence to another agency and receipt of submissions for analysis.* If the police or scene of crime officers are submitting the evidence collected from a scene to a laboratory for analysis, they must take it to a submissions office or similar place. At this time, the items will be logged and a reference number assigned by the submissions office or the laboratory that is taking receipt of the items.

- *Choice of analysis.* In many cases, a limited sample is available and it is not possible to return to the scene to collect more samples; therefore, the choice of the analysis type is crucial. If the wrong type of analysis is carried out, we may have compromised any other analyses that can be conducted or we may use the sample up so that no further testing can be carried out with the data obtained. This is why it is necessary to have standard operating procedures in order to ensure that we get the best out of the sample with which we have to work.

- *Integrity of data.* It is necessary to ensure that all records and reports correlate with the raw analytical data obtained from analyses. This is to make sure that analysts have transferred the required results from the raw data to the records or report. It is also used to ensure that analysts have not fabricated data to fit with a specific outcome because both of these could greatly affect the relationship of the organisation with clients and may also harm public opinion.

- *Statements.* Having a standard format for statements makes reporting work for court easier. The format could vary greatly from one individual to another. Some may only include minimal information, whereas others may wish to include all raw data from their analyses. This is unhelpful for the jurors and other court officials required to read the output and it may hinder rather than help.

## Introduction to Quality Assurance

Quality assurance is a planned and systematic pattern of all actions necessary to provide confidence that adequate technical requirements are established, that products and services conform to established technical requirements, and that satisfactory performance is achieved. This means that quality assurance is proving that what we do is fit for purpose.

There are a number of reasons for introducing quality assurance into a forensic science laboratory. One would be to ensure that we avoid issues in relation to the product (evidence) that we are analysing that might result in inappropriate analysis and destruction of the item or evidence. Another reason for quality assurance is to ensure customer (police and judicial systems) trust, and the last would be to avoid miscarriages of justice (i.e., that guilty individuals do not walk free and innocent individuals are not given a prison sentence).

## How Do We Ensure Quality in Our Laboratories?

In order to ensure quality in our laboratories, we can introduce quality procedures. The following factors must be considered and action implemented:

- *Documentation.* The documentation involved in a quality management system includes a quality manual and a standard operating procedure (SOP) manual. A quality manual is concerned with all aspects of the administration of an organisation and should include the management structure, job specifications, staff training records, and appraisal systems in place. The SOP manual, on the other hand, should contain detailed instructions for each task carried out within the organisation. Each procedure must be documented and each step of the process thoroughly explained so that new members of staff can carry out the procedure to the same standard using the SOP.

- *Document control.* A unique numbering system is used that will include information on any revisions that have taken place in order to control the documents. This control means that only a limited number of copies of the document will be allowed in circulation in the organisation (photocopying is prohibited). The documents will have to be released officially and signed and dated by the individuals who wrote it and by a senior member of staff who controls all of the documents. (This is usually done by a quality manager whose sole responsibility is to ensure that the quality management system in place in the organisation is conformed to and oversees the whole process.)

- *Staff training.* Staff training and the recording of this training are essential in a quality assured laboratory environment. In order for an individual to be compliant with the quality system in place, he or she must be fully aware of what is involved in the work of the organisation; this means that individuals are required to undergo training in the work that they are to carry out and also to read and understand the SOPs that they will have to use. All of this should be documented in order to be able to supply clients and regulating bodies with evidence that all staff are adequately trained and that they have signed to show that they have read and understood the appropriate SOPs. Staff should also be encouraged to undergo continual professional development, and any courses or training attended should be included in the individual's training record (along with the initial staff training and signed SOP read lists). SOP updates, when made available, should be read by the staff member and that fact added to his or her record when appropriate.

- *Instrument qualification.* This step is a means of measuring an instrument's quality and ensures that we have consistent instrument performance. This is of prime importance in forensic science analyses because we need to be sure that the results that we are reporting are consistent. For example, consider that we have obtained a value for the presence of morphine in blood that is indicative of a lethal concentration; later that day, we run the sample again, on the same instrument, and find that the value we have determined on this occasion is half that of the original. This is a very extreme scenario; nonetheless, we must be able to show that the instrument we are using for our analyses will give us the same results each time we carry out the work (an acceptable level of error will usually be associated with the value that we report and be outlined in an SOP). A number of steps are involved in instrument qualification: design qualification (DQ), operation qualification (OQ), and performance qualification (PQ); these are explained in Chapter 8.

- *Validated testing procedures.* We must be able to show that our testing methods are fit for purpose to ensure that we have integrity of the analytical data. To do this, we must have validated methods that conform to guidelines set out by accrediting bodies or organisations such as the International Conference on Harmonisation (ICH) or Eurachem. Here, parameters are defined to validate an analytical testing procedure and acceptable ranges provided, where appropriate. If, one day a blood sample is submitted for a general unknown screen and we carry out the test using one validated method and the next day we do the same test on a different blood sample and we alter the method, there will be no consistency in our results. We will no longer be sure that the method is specific for all of the compounds that it was previously possible to detect.
- *Dealing with errors.* A number of different types of errors can occur and are covered in Chapter 10. Here we consider what we should do when these errors occur. First and foremost, we must always admit the error and never try to cover it up because it will always return and the repercussions could be immense. We must inform a more senior member of staff and be prepared to repeat work if possible or necessary. This obviously involves usually expensive retesting and can result in a termination of a contract; more importantly, it may result in a loss of public confidence.

## A Quality Management System

Making the decision to implement a quality management system (QMS) is only the first step of the process of having a documented quality system in place in the workplace. The next step in the process will be to decide which accreditation body best suits the needs of the organisation. The quality management system comprises:

- The *policy document* defines the aims of the accreditation relative to the company; the rest of the quality management system will be based on this documentation.
- The *quality manual* outlines the policy statement, the roles and responsibilities of the employees and management, and the procedures involved in the running of the organisation.
- The documents included in the *procedures manual* are all of the SOPs that relate to specific activities, methods, and instrumental techniques within the organisation. They are set out such that they provide step-by-step instructions for use.

- *Raw data* include anything obtained from instruments, anything written on (e.g., completed fridge or freezer logs), and reports released to customers.

The benefits of having a quality management system are that we have a form of internal quality control that allows us to have consistency of the service we provide. Our work will be verified, documents checked, and internal quality audits conducted. This means that we are continually assessing the practices set down in the documents of the quality management system.

## Accreditation

The process of accreditation (Table 9.1) starts with obtaining a copy of the documentation from the accrediting body that will outline the standards and criteria to be met. Next, the quality and SOP manuals should be written (or altered if they are already in use) according to the guidelines of the regulating body. From this point, the documents and other appropriate paperwork will be sent, usually with payment of a fee, with the initial application.

There are many accrediting bodies across the world, such as the United Kingdom Accreditation Service, or UKAS (UK); American Association of Crime Laboratory Directors/Laboratory Accreditation Board, or ASCLD/ LAB (United States); and the National Association of Testing Authorities, or NATA (Australia). The process will be specific to each accrediting body, but in general an application will be submitted whereby an assessment manager will be assigned. This assessment manager usually deals with the organisation from the initial application throughout the rest of the process of accreditation. In this process the organisation will be subject to a preassessment visit, an initial assessment visit, and visits to maintain the accreditation, known as audits.

**Table 9.1   Disadvantages and Advantages of Accreditation**

| Disadvantages | Advantages |
| --- | --- |
| Emphasis on regulation | Pride from accreditation |
| More records to keep; more work | Quality of product or service should be more consistent |
| Cost of the QMS and staff involved | Methods and instructions are documented |
| Difficulties in asking suppliers to conform to the system chosen or finding new suppliers | Clear roles and responsibilities are outlined |
| Procedures | Improves overall process |
| | Documentation system will be in place |

**Table 9.2    Types of Audits**

| Audit Type | Carried Out by | Description |
|---|---|---|
| Internal | Organisation | First party |
| External | Organisation or a supplier | Second party |
| Independent | Accrediting body | Third party |

As part of most, if not all, accreditation programmes, laboratories must participate in a proficiency testing programme in which all of the laboratories involved in the programme are sent a sample containing the same amounts of substances. Each lab will carry out its work and return its findings to a central laboratory or other organisation that will collate all of the data and return them to each participating laboratory. This returned report will include the overall scoring of the laboratory in relation to the sample prepared and also in relation to the other laboratories that participated.

## Auditing

The purpose of completing an audit is to check the effectiveness of the systems in place within the organisation. From the audits carried out, problems and potential problems can be identified, thus leading to changes that will improve the overall system. An audit is defined as 'a systematic, independent and documented process for obtaining audit evidence and evaluating its objectivity to determine the extent to which the audit criteria are fulfilled' (BS EN ISO 19011:2002). Table 9.2 shows types of audits.

Audits should be scheduled (although unannounced audits can be carried out), preplanned, structured, independent, objective, competent, completed on time, and reported. Any noncompliances found by the audits will be reported and a review of the processes and procedures will be carried out and any amendments made.

---

**KEY POINT SUMMARY**

- Quality assurance is a planned and systematic pattern of all actions necessary to provide confidence that adequate technical requirements are established, that products and services conform to established technical requirements, and that satisfactory performance is achieved.
- Accreditation is implementing a quality system through an accrediting body, such as UKAS, ASCLD/LAB, and NATA.
- Auditing is used to check the effectiveness of the systems in place within the organisation.

## Further Reading

Hibbert, D. B. 2007. *Quality assurance in the analytical chemistry laboratory.* New York: Oxford University Press.

Pritchard, E., and V. Barwick. 2008. *Quality assurance in analytical chemistry.* New York: Wiley-Interscience.

Taylor, J. K. 1987. *Quality assurance of chemical measurements.* Boca Raton, FL: CRC Press.

www.ascld-lab.org

www.iso.org

www.nata.asn.au

www.ukas.com

# Troubleshooting HPLC Systems

# 10

## Introduction

In HPLC it is inevitable that something will break down at some time because so many interactions are taking place within the chromatographic process—not to mention the many moving parts within the HPLC instrument. It is tempting to dive in and try to fix the problem. However, a word of caution: You might do more harm than good. Every action needs to be documented and monitored carefully to ensure quality compliance and to ensure that future results are reliable and accurate.

Before undertaking any troubleshooting operation, ensure that you have the permission of the person in change of the HPLC system. It is advisable for a novice to carry out the work under the supervision of a more experienced analyst or technician. We would imagine that, for undergraduate students, most of what follows is probably done for you by your university laboratory technicians. It is important, however, that you understand what is being done behind the scenes. As you take on more responsibility as part of your job, a postgraduate project, or even an undergraduate project, you will find that many things can go wrong with an HPLC system.

Although the reasons can often seem to be complex in nature, a systematic approach to problem solving is the best way forward. Quite often, the simplest things cause the biggest problems and can prove time consuming to fix. Troubleshooting in HPLC has been the subject of many books and literature guides produced by a variety of individuals and HPLC providers. Lots of information is available through the Internet and some of the best of this information is provided in the 'Further Reading' section of this book.

The scope of this chapter is limited to troubleshooting HPLC systems using predeveloped and validated methods of analysis. We will focus on some of the most commonly encountered problems and the possible solutions.

In order to understand the problem, it is necessary to understand how each of the components functions in the HPLC system and the chromatographic processes that are taking place. The main components that make up the HPLC system have been explained in Chapter 2, namely:

- pump
- injector

- column
- detector

While one is engaged in troubleshooting, it is necessary to consider the source of the problem and to examine how the problem might have arisen. Additional problems might arise from the sample composition and the mobile phase, and these also need to be considered. It should be noted that problems can present themselves in many different ways, and what might appear to be a pump-related problem can quite easily be a result of something else entirely.

Different authors approach troubleshooting in different ways. Some prefer to present symptoms and solutions and others prefer to look at component-related problems and provide solutions. It is difficult to know which way is better, but we have always felt that the best source of information in relation to what is going on is the chromatogram. The chromatogram is the pictorial representation of what has happened in the system and therefore is a measure of what is right and what is wrong—for example:

- Is there a peak where you would expect to see one?
- What does the baseline look like?
- Is the retention time stable?

## Troubleshooting

Troubleshooting comes with practice and, as you gain experience, the techniques of troubleshooting will become easier. Being able to spot when something is not right is half the battle. A good chromatographer should always be asking questions, and we have provided a question approach to troubleshooting. It aims to identify the source of a problem quickly, primarily through the examination of the chromatogram and the following questions:

1. Is there any signal?
2. Is there a peak?
3. Are the retention times as expected and within established tolerance limits?
4. Does the baseline look unusual; for example, is there a lot of baseline noise?
5. Is the resolution as expected and within established tolerance limits for the peaks of interest?
6. Are the peaks a good shape (Gaussian)?
7. Does everything sound OK?
8. Is the system pressure OK?

These questions should provide a starting point for any troubleshooting activity and should be a matter of course when assessing any chromatographic procedure. Taking the symptom/solution approach, we can look at each of the questions and assess where problems might arise, how they can be diagnosed, and how they can be solved.

### Question 1: Is There Any Signal?

There is no signal.

### *Possible Causes*

- The detector is not switched on (see Figure 10.1).
- The bulb in the lamp has blown.

### *Diagnosis and Solutions*

Not switching on the detector is a basic error that is encountered more than might initially be thought. This can be solved quite easily by switching on the detector and allowing the lamp to warm up for about 20 minutes. This warm-up time will vary from manufacturer to manufacturer and it is always advisable to check with the manufacturer's documentation before running any samples. If the lamp bulb is blown, then a new one should be fitted. This procedure is typically straightforward on modern HPLC systems. It is usually just a case of slotting the new lamp into position and tightening a couple

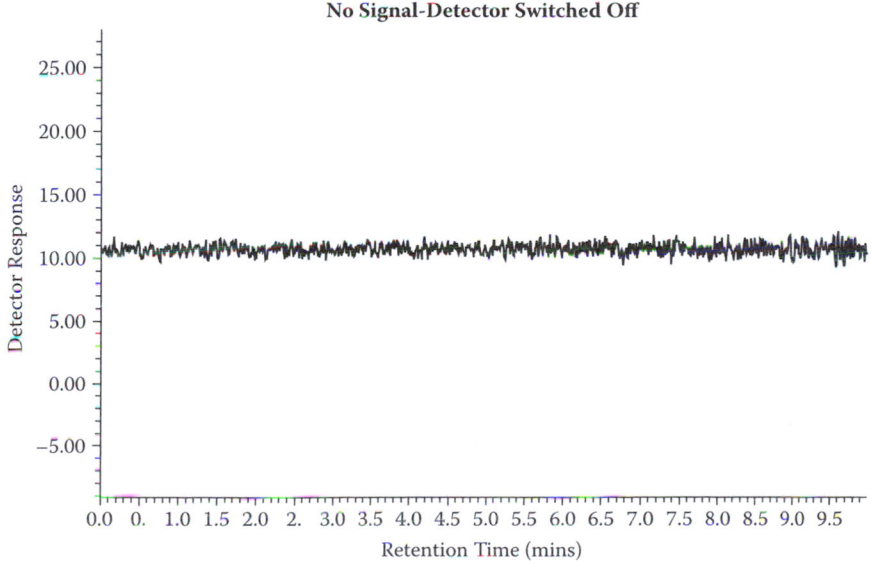

**Figure 10.1** Detector switched off.

of screws. HPLC systems are designed to ensure that the lamp cannot be inserted in the wrong orientation. The lamp timer should be reset to zero and system performance qualification (see Chapter 8) conducted before any further samples are run.

### Question 2: Is There a Peak?

There is no peak, but I can see a baseline signal (see Figure 10.2).

#### *Possible Causes*

- The sample does not contain the compound of interest at a high enough concentration to be detected under current conditions.
- There is a blockage in the injector.
- The desired vial is not in the correct position in the sample tray.
- The pump is not switched on or is not pumping.

#### *Diagnosis and Solutions*

- Check the sample preparation steps (see Chapter 3) and prepare the sample solution again at the recommended concentration (if necessary) and reinject.
- Clear any blockage in the injector (refer to the instrument user manual).

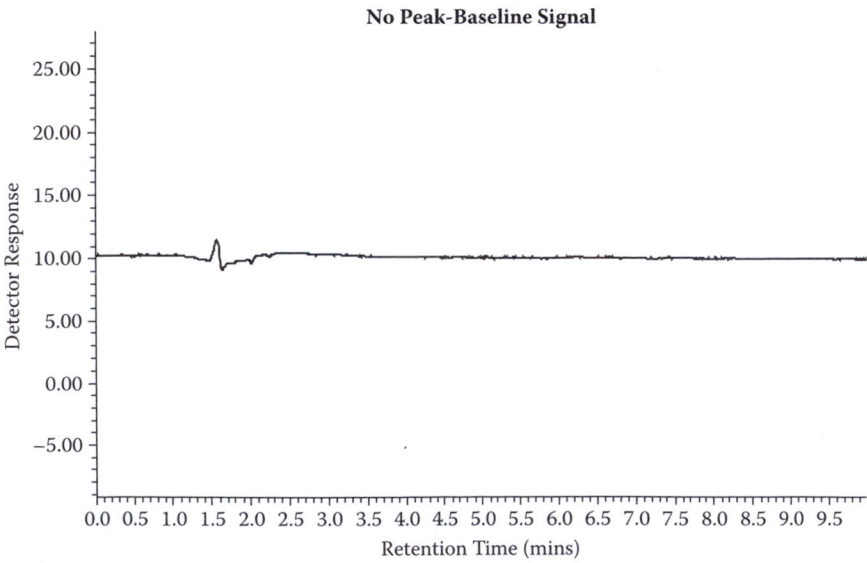

**Figure 10.2**  Detector on—no peak.

- Needle: common causes include deposited buffer salt or rubber from vial septum.
- Valve: common causes include deposited salt or broken motor valve.
- Syringe barrel and plunger: this is more commonly encountered with manual injection systems and situations where the sample matrix is 'sticky'. The barrel and plunger should be washed with aqueous solvent to remove the sticky material.

- Ensure that the desired vial is in the correct position and reposition if required. This is a common and easy mistake to make.
- Switch on the pump and prime with mobile phase to remove any air by following the manufacturer's instructions.
  - Ensure that there is sufficient mobile phase within the solvent reservoirs. The mobile phase may have run out, causing the pump to shut down.
  - Inspect the pump head for damage; for example, check the pistons and valves and replace any faulty parts as required. Always refer to the manufacturer's instructions when changing a part (see Chapter 8).

### Question 3: Are the Retention Times as Expected and within Established Tolerance Limits?

The retention time is not as expected—either shorter (more usual) or longer than usual.

### Possible Causes

- There is a leak somewhere in the system.
- There is a problem with the mobile phase. This can result in a decrease or an increase in retention time and can usually be attributed to one of the following:
  - incorrect mixing (Figure 10.3)

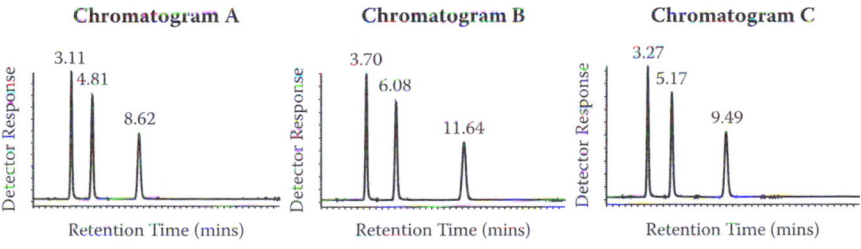

**Figure 10.3** Representation of manual mixing errors.

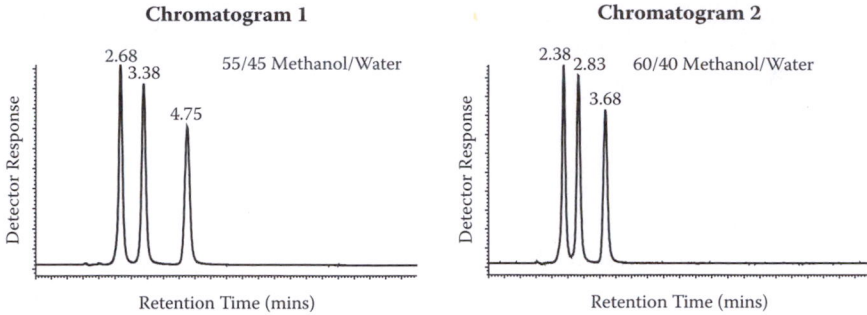

**Figure 10.4** Effect of a small change in organic modifier.

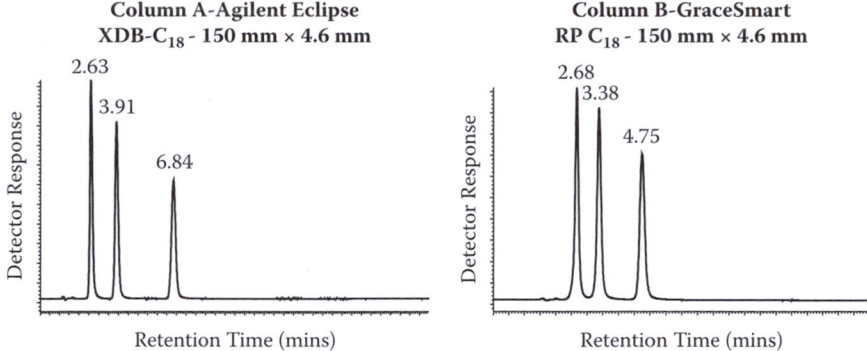

**Figure 10.5** Incorrect column.

- incorrect pH
- evaporation of organic phase (Figure 10.4)
- There is some problem or issue associated with the column (Figure 10.5).
- Temperature (Figure 10.7).

## *Diagnosis and Solutions*

Evaluate and measure the void time ($t_m$; time for elution of an unretained peak; see Chapter 2) and compare this with previous injections. If there has been a change in the void time, then the problem is associated with a change in flow rate, which might be the result of a leak in the system. If the void time has not changed, then the problem is likely to be associated with the mobile phase, the stationary phase, or the temperature. Another thing to consider here is that the column packing may be dissolving and the retention characteristics of the compounds of interest are no longer maintained (see Chapter 6).

1. *Leaks.* Check the instrumentation in a systematic manner in search of the leak. It is useful to check the flow rate at this point by placing a measuring

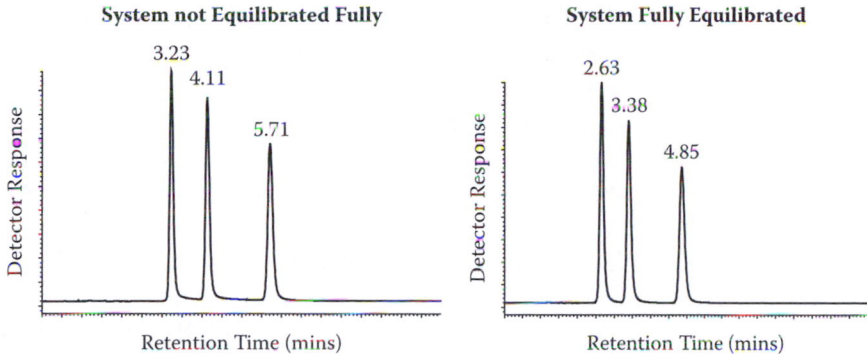

**Figure 10.6** Column equilibration—column C18, 150 mm × 4.6 mm, 5 µm; mobile phase 60/40 methanol/water.

cylinder under the mobile phase outlet and measuring the amount of mobile phase dispensed in a 10-minute time period with the aid of a stop watch. Leaks can occur anywhere within the tubing network, but they occur most frequently around the column connectors as a result of poor installation. This is an ideal starting point from which to begin the search and the column should be removed and refitted, ensuring that there is a good seal around the fittings. The fittings may need to be replaced.

**Note:** Before disconnecting any of the tubing, ensure that the pump is switched off. A sudden release in system pressure may result in a fine, high-pressure jet of mobile phase, which can pose health and safety risks.

If reconnecting the column fails to correct the leak, then other common areas to search for leaks are the detector flow cell, the pump (around the pistons), and the injector valve. The same process should be followed until the source of the leak is identified and corrected. New parts may be required; for example, the pump piston seals may have worn out. Always consult the manufacturer's user manual when attempting to replace any part. If in doubt, contact the manufacturer and ask for installation of the new part. It will be necessary to undertake system performance qualification on the replacement of any worn out or damaged part (see Chapter 8).

Some HPLC systems have leak detectors placed at strategic locations on the system, which can be invaluable in this sort of situation. A leak on these systems is often accompanied by a red warning light and a pump shutdown. Although quite obvious, another useful diagnostic tool is to look for a small deposit of mobile phase around or under the HPLC system. It is easier to see this when the area around the HPLC is kept clean and tidy, and good house-keeping practice at all times is recommended.

Small pools of mobile phase can be a common occurrence when buffers are used in the mobile phase. Where there are high concentrations of buffer and/or high levels of organic modifier, precipitation of buffer salt is common around the piston area on the pump. Constant rubbing of the piston against the seal in the presence of salt deposits causes the seals to wear away and ultimately to leak. Many HPLC systems come equipped with a seal wash, which ensures that any buffer deposits are flushed away during the course of the chromatographic process. This ensures that deposits are minimised. Seal wash reservoirs must always be topped up with an appropriate solvent compatible with the mobile phase. It is not recommended to fill the solvent reservoir with 100% organic phase.

2. *Mobile phase.* Many of the problems associated with the mobile phase will depend on the HPLC instrument configuration and, as a consequence, whether the mobile phase preparation is manual or automated.

**Mixing errors: manual mobile phase preparation.** These tend to occur as a result of operator error during the dispensing and mixing of the individual components. The final concentration of each component within the mixture can vary depending on the way in which it is prepared.

### HYPOTHETICAL EXAMPLE 10.1

You have been asked to prepare 1 L of a 60/40 methanol/water mobile phase for use in an assay of basic compounds on an ODS type of column.

Method 1: add 400 mL of water into a 1 L volumetric flask and make to volume with methanol. This will result in a solution that contains too much methanol and the retention times for basic species will shorten (see chromatogram A in Figure 10.3).

Method 2: add 600 mL methanol into a 1 L volumetric flask and make to volume with water. This will result in a solution that contains too much water and the retention time for basic species will lengthen (see chromatogram B in Figure 10.3). An explanation of why this occurs is given in Chapter 6.

Method 3: measure 600 mL of water and 400 mL methanol individually and mix together in an HPLC solvent bottle. The resulting solution will contain the correct amount of each phase (see chromatogram C in Figure 10.3).

**Mixing errors: automated mobile phase preparation.** Problems encountered with an automated system can still be the result of analyst error. For example, the wrong component may have been dispensed into the HPLC solvent reservoir. Where there is an apparent issue in relation to the retention time of a compound, the mobile phase should always be checked for accuracy of preparation. Proportioning valve errors can also cause a problem with automated systems. When the valve ceases to dispense the correct amount of either solvent, the retention time of the compounds of interest will vary.

**Proportioning valve errors** can be intermittent and this may lead to inter- and intrarun variations in retention times. If it is suspected that there is a problem with the proportioning valve, the manufacturer should be contacted and an arrangement made to repair the system. This type of maintenance would not be considered routine and, as such, should not be attempted by an inexperienced analyst. Undertaking this type of activity may invalidate any manufacturer's warranty or maintenance agreement.

**Errors with pH.** As discussed in Chapter 3, an incorrect mobile phase pH can have a detrimental effect on the retention time of a compound. Ionisation of a compound makes the compound less hydrophobic; therefore, the affinity for the stationary phase will be reduced and a change in the retention time will take place. During the method development and validation stages, the method will have been developed to allow for subtle changes in the mobile phase composition in relation to pH; however, if the wrong buffer or a significant error has been made in the preparation, more pronounced shifts will be experienced.

**Evaporation of the organic modifier.** This problem is usually restricted to mobile phase mixtures that have been prepared manually. When mobile phase mixtures are stored, there is an air space above the level of the liquid. This air space will increase as the mobile phase is used up during the run. If the run time (total time taken to run all samples) is long or the solvents are in use over several days, the organic portion of the mobile phase can evaporate off into the air space, thus altering the mobile phase composition and hence retention times (see Figure 10.4). This problem can be overcome to a certain extent by ensuring that the mobile phase is stored in a suitably sized container with a narrow neck and by ensuring that the container is sealed. This can be performed by using a screw cap or by applying Parafilm® (or suitable alternative) over the container neck. Mobile phases should be prepared regularly and should be assigned an expiry date.

3. *Column.* Problems with the HPLC column are due, in the main, to a small number of possibilities:

- Column choice was incorrect.
- The column has deteriorated beyond its useful lifetime and the packing material is no longer retaining compounds to an acceptable level.
- Insufficient time has been allowed to ensure complete column equilibration.
- Unacceptable levels of contamination were present.

Dealing with these problems is usually straightforward: *Change the column.*

Incorrect column choice is usually due to analyst error (see Figure 10.5). Where there are significant changes in the chromatography, the column specification should be carefully checked and the column replaced if necessary. Column information is printed on the barrel of the column.

Columns deteriorate with use over time. Extremes of pH and constant use can cause the packing material to be stripped away from the silica backbone. This causes silanol sites, which interact with components within the analytical sample, to be exposed on the surface of the packing material. Silanol sites are acidic in nature and, at normal operating pH, these will become negatively charged and will interact with any positively charged basic components within the sample mixture. This will result in peak tailing (see Chapter 2).

As the number of active hydrophobic sites is reduced, the interaction with nonpolar species within the sample becomes less strong and retention times shorten. The capacity of the column has been reduced—hence, shorter retention times. As the capacity is reduced, peak tailing occurs due to a band broadening type of effect. Problems of this nature can be prevented before they happen by ensuring that there is a system suitability section within all method SOPs (see Chapter 8). System suitability assesses parameters that indicate whether the column is performing to an acceptable level and measures such parameters as theoretical plates, tailing factor, retention time, and resolution. Setting limits around each of the parameters alerts the analyst early to any problems that might have arisen before too much time has been wasted.

Insufficient conditioning of the column with mobile phase will result in retention time drift (see Figure 10.6). If the amount of time allowed for the conditioning process is too short, there will be a shift in retention time over the course of a large number of samples as the column continues to equilibrate. The column equilibration time will be dependent on the dimensions of the column. It is usual to condition the column with between 10 and 20 column volumes of mobile phase (see Table 10.1).

4. *Temperature.* Fluctuations in temperature can be associated with two things: the environment in which the HPLC instrument sits and instrument error. As we know, there are temperature fluctuations (see Figure 10.7) in the external environment as we go through the day and into the night or as we move through the seasons. If we do not control these fluctuations in temperature indoors, the HPLC retention times will drift with changes in ambient temperature. For example, an unregulated (no air-conditioning) indoor environment will increase in temperature as the day progresses and will

**Table 10.1   Column Equilibration Times (Approximate Values)**

| Column Dimensions (mm) | Column Volume (mL) | Flow Rate (mL/min) | Equilibration Time (min)[a] |
|---|---|---|---|
| 250 mm × 4.6 mm | 2.90 | 1.0 | 58 |
| 150 mm × 4.6 mm | 1.75 | 1.0 | 35 |
| 100 mm × 4.6 mm | 1.16 | 1.0 | 23 |
| 150 mm × 2.1 mm | 0.37 | 0.2 | 37 |
| 100 mm × 2.1 mm | 0.24 | 0.2 | 24 |

[a]  Based on 20 column volumes.

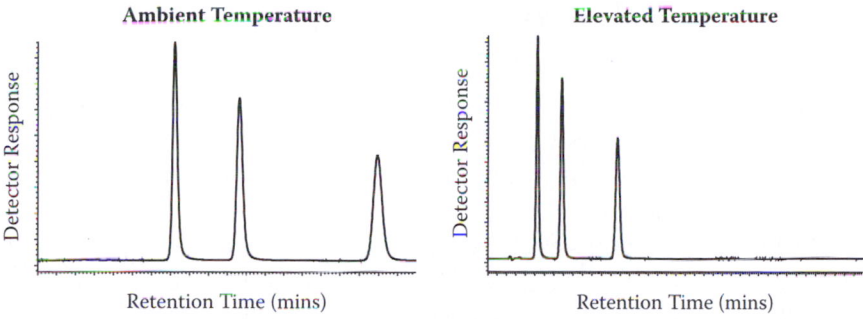

**Figure 10.7** Influence of temperature on retention time.

reduce in temperature again during the night. Some companies switch off air-conditioning systems at night in an effort to save money. It is worthwhile investigating company policy in relation to this so that potential problems can be avoided.

Temperature can be regulated with the use of a column compartment heater (typical temperature range of 25–80°C) or the air-conditioning can be switched on locally. An increase in temperature reduces the mobile phase viscosity and improves the mass transfer, and compounds elute at a faster rate. This can result in loss of resolution between compounds and can have a detrimental effect on the assay outcomes. A decrease in temperature will have the opposite effect, but can lead to problems nevertheless. It is usual to run HPLC systems at a temperature slightly above ambient (25°C) when a column compartment heater is used.

## Question 4: Does the Baseline Look Unusual and Is There a Lot of Baseline Noise?

There is a lot of baseline noise.

### *Possible Causes*

- air in the detector flow cell
- contamination
- lamp deterioration

### *Diagnosis and Solutions*

Baseline problems in isocratic systems can usually be spotted due to spikes or to waves. Spikes usually occur as a result of air trapped in the detector cell and waves usually occur as a result of lamp deterioration or contamination. A good place to start with any baseline issues is to switch off the pump and continue to monitor the baseline.

**Note:** Most systems have the facility to monitor the baseline while the system is not pumping in the mobile phase. Check with your system provider.

If the problem disappears, then it is reasonable to expect that the problem lies with unwanted contamination or air in the detector flow cell. If the problem remains, it is reasonable to expect that the problem is associated with lamp deterioration.

Air in the detector flow cell can be gently removed from the system by increasing the mobile phase flow rate slightly. This increases the pressure, forcing the air bubbles out of the flow cell and back into solution. One way to avoid air in the detector flow cell is to maintain a positive pressure at the detector outlet. This can be achieved by attaching a short coil of low-diameter HPLC tubing to the flow outlet. This acts to increase the pressure and ensures that air remains dissolved in the mobile phase. Thorough degassing of the mobile phase is recommended before use in order to avoid these problems.

**Note:** Care must be taken not to exceed the column operating pressure or to exceed the maximum working pressure for the flow cell.

Contamination can be the result of a buildup of retained sample material on the stationary phase. This can be a problem with gradient elution systems when the proportion of organic modifier in the mobile phase increases over the course of the chromatographic run. These unwanted materials can be washed off the column with the increase in mobile phase strength, which in turn can distort the baseline in the form of waves. In some circumstances, this can interfere with late eluting peaks of interest and can have a detrimental effect on the assay outcomes.

Contamination or dirt can also accumulate at the top of the HPLC column in the packing material or in the retaining frit. As the chromatographic process progresses, the dirt is gradually washed through the system, appearing as waves on the baseline. This problem can be solved by washing the column with a high proportion of organic modifier in the mobile phase. This assumes that the dirt is soluble in such a mobile phase and that it can be removed easily. One way to avoid contamination issues of this sort is to use a guard column filled with the same packing material or a compatible packing material. Ensuring that both the mobile phase and the sample are filtered before use also helps to eliminate contamination issues of this nature. Always use HPLC-grade solvents where appropriate and affordable and ensure that water supplies are of a suitable quality.

A dirty column can be cleaned out in one of two ways: by removing the dirty packing material or by removing the top of the HPLC column and scraping out a small amount of material. If the contamination is deep, it is often advisable to change the column at this stage. Another remedy is to remove the column from the system and reattach it the opposite way round.

Packing materials are packed at a high enough pressure to withstand mobile phase pumping against the direction of flow.

Just as the bulb in a table lamp has a finite lifetime, so do the ultraviolet/visible (UV/Vis) detector lamps (tungsten and deuterium; see Chapter 4) in HPLC systems. Light intensity decreases with time due to the evaporation of internal metal components and coatings. Lamps deteriorate with age and use; according to most manufacturers, lamps have a lifetime of about 1,000 hours. The lifetime is defined as being the time at which 50% of the light intensity remains. After this time has elapsed, the lamp should be replaced.

Storing lamps for long periods of time is not recommended because prolonged storage can reduce the expected lifetime significantly. Leaving the lamp on for long periods of time or frequently switching the lamp on and off will adversely affect the lamp lifetime. Monitoring the lamp performance should be carried out routinely and the lamp replaced once a certain threshold value has been reached. This can be measured in hours of use or in lamp output, depending on the system used. Check to see what the system manufacturer recommends.

With gradient elution HPLC, a drifting baseline is commonplace due to changes in the mobile phase absorbance as the proportions change within the mobile phase composition. This effect can be reduced by using solvents specifically designed for use in gradient elution or by using solvents with closely matching UV absorbance.

## Question 5: Is the Resolution as Expected and within Given Tolerance Limits for the Peaks of Interest?

The resolution is not as expected and fails to meet the given tolerance limits for the assay.

### *Possible Causes*

Problems with resolution are usually caused by the same issues that result in changes in retention time. These are listed again as follows. Reference should be made to the previous section for further information.

- There is a leak somewhere in the system.
- There is a problem with the mobile phase. This can result in a decrease or an increase in retention time and can usually be attributed to one of the following:
  - incorrect mixing (see Figure 10.3)
  - incorrect pH
  - evaporation of organic phase (Figure 10.4)
- There is some problem or issue associated with the column (Figure 10.5).

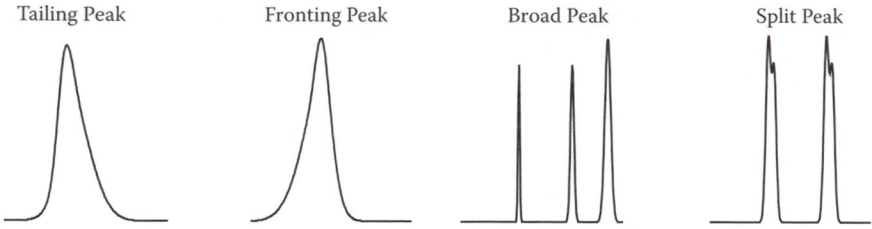

**Figure 10.8** Illustration of peak shape.

## Question 6: Are the Peaks a Good Shape?

Peak shape is poor (see Figure 10.8). Poor peak shape can take different forms and includes peak tailing, peak fronting, broad peaks, and peak splitting. Each of these different forms will be dealt with individually.

### *Peak Tailing: Possible Causes*

- incorrect mobile phase—pH
- incorrect column
- column void
- coeluting peak
- column has deteriorated beyond its useful lifetime

### *Peak Tailing: Diagnosis and Solutions*

Acceptable levels of peak tailing should be indicated in the system suitability section of the method as *peak asymmetry* or *tailing factor*. Once the limits have been exceeded, a number of basic checks can be carried out that can assist in the diagnosis. A lot of problems of this nature are found to be due to analyst error, and the mobile phase composition and column choice should always be checked as a matter of course.

**Note:** In the case of an incorrect column, the wrong one should be washed out thoroughly with an appropriate solvent before storage.

If either the column or the mobile phase is incorrect, then this should be put right immediately and the system allowed to reequilibrate under the correct operating conditions before repeating the analysis.

**Note:** Before repeating any analysis, always consult with your supervisor or line manager and ensure that you are following the correct procedure and protocol for such eventualities.

Voids in the column packing material can be harder to spot, depending on the severity of the void (see Figure 10.9). Dismantling the column should only be performed as a last resort. Opening up the column can disturb the

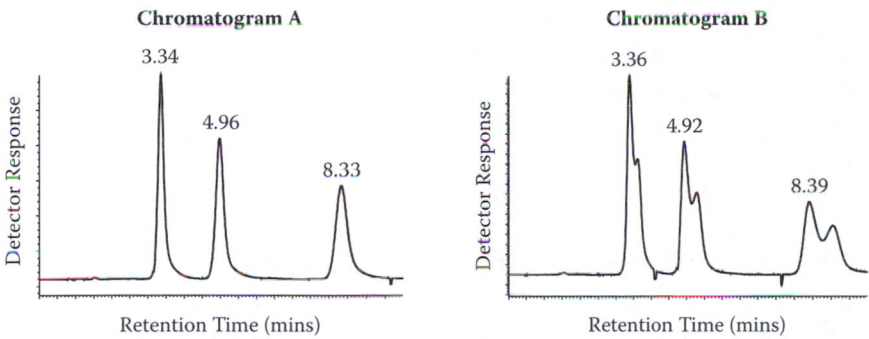

**Figure 10.9** Chromatogram A represents a small column void; chromatogram B represents a much larger column void.

packed bed, causing further damage. If a void is found, then this can be topped up using a slurry made up of the packing material suspended in an appropriate solvent (usually methanol). The slurry should be applied in small amounts and the solvent allowed to soak into the bed before applying the next aliquot.

The column should be reassembled and reattached to the HPLC system. It should be allowed to reequilibrate and an injection should be performed. The peak shape should be assessed and, if an improvement is apparent and the system suitability criteria are met, the column can be used for the analysis. Quite often it is necessary to repeat the slurry application several times in order to fill the void completely. This can be time consuming and in some instances the preferred option is to replace the column.

Coeluting peaks can occur as a result of a deteriorated column. As discussed earlier, the retention time of a compound can shift and, when this occurs, peaks may appear to coelute partially. The interaction between a basic compound and the exposed active negative sites on the column can give the appearance of a peak tail (see previous discussion). This can normally be rectified by changing the column.

### *Peak Fronting: Possible Causes*

- The sample solvent is incompatible with the mobile phase.
- There is sample overload.

Peak fronting normally occurs because of an incompatibility between the sample solvent and the mobile phase or as a result of the sample being too concentrated. If a sample is prepared in 100% methanol and the mobile phase contains 5% methanol, then this might result in a fronting problem. Under normal conditions, when the sample is introduced into the flow of mobile phase, the sample solvent and the mobile phase quickly mix. The sample then begins the journey through the column in keeping with its affinity for the mobile phase or the stationary phase. In cases where there is a mismatch between the sample solvent and the mobile phase, the strong sample solvent does not immediately mix with the mobile phase and the solvent and stationary phase will compete for the sample. This means that some of the sample will reside in the solvent and pass quickly through the column, leading to a fronting peak.

Too much sample means that the capacity of the column will be exceeded and some of the compound cannot interact with the stationary phase. The material that does not interact with the stationary phase will travel at a faster rate through the column, resulting in peak fronting.

### Peak Fronting: Diagnosis and Solutions

Issues in relation to sample solvent incompatibility can be overcome by changing the sample solvent to one that is compatible with the mobile phase. Care must be taken to ensure that the sample is soluble in the solvent of choice at the levels required for the assay. When the column capacity has been exceeded, there are two possible solutions: change the column or dilute the sample. A column packed with a higher capacity stationary phase might suffice, but it may be necessary to change the column dimensions as well. Changes in concentration may cause issues if the analyte of interest is present in low concentration in relation to a much larger impurity. In this case, it may be necessary to perform an extraction process such as liquid–liquid extraction (LLE) or solid phase extraction (SPE) to remove the high-level impurity.

### Broad Peaks: Possible Causes

- The injection volume is too large.
- Column efficiency is low.
- The retention times are too long.
- There are late eluting peaks in the chromatogram (carryover).

### Broad Peaks: Diagnosis and Solutions

If the injection volume is too large, this can be reduced as long as the solution is sufficiently concentrated and will remain within the validated range of the assay. If this is not the case, then it may be possible to concentrate the solution using one of the extraction techniques discussed in Chapter 3. Low column efficiency is probably due to the column being old and it can be overcome by replacing the column with a new one. If the retention times are too long, this would suggest that the wrong column has been used. The column details should be checked and the column replaced where necessary. The same applies to late eluting peaks that are the result of carryover. This would suggest that the method has been set up incorrectly on the system and that insufficient time has been allowed for all of the components to elute from the column. The method parameters should be checked and adjusted where necessary.

### Splitting Peaks: Possible Causes

- contamination
- injector error
- sample solvent incompatible with the mobile phase (see peak fronting)
- column void (see 'Peak Tailing' and Figure 10.8)

### Splitting Peaks: Diagnosis and Solutions

Contamination in the system is usually the result of a buildup of unwanted material at the column inlet, either in the packing material or on the column frit. This can often be removed by back-flushing the column or by replacing a small amount of the packing material as required. Problems with the injector are usually the result of a worn or scratched stator valve. If the face becomes scratched, the sample can leak into the scratched area; this causes the sample plug to be split at the point of injection and hence the split peak.

## Question 7: Does Everything Sound OK?

There is a strange noise coming from the HPLC system.

### Possible Cause

This is usually indicative of a broken mechanical part within the system.

### Diagnosis and Solutions

Unless you are a very experienced operator, it is best to leave the replacement of moving parts to the system manufacturer. Try to isolate the broken part and then call your engineer.

## Question 8: Is the System Pressure OK?

The system pressure is higher or lower than normal or is fluctuating up and down.

It is always advisable to record the system operating pressure at the start of every HPLC sequence or at the beginning of every day in order to establish a baseline or reference for each combination of assay, column, and system. If the operating pressure is considerably higher or lower on any occasion, this usually indicates that something is not right. Sudden increases in pressure are usually due to issues with the column or mobile phase, and sudden drops in pressure are usually the result of a small leak in the system or a change in the flow rate.

## Possible Causes

Increases in pressure can be the result of

- a buildup in particulate matter in the column or in the HPLC tubing
- precipitation of buffer salts
- an incorrect column (analyst error)
- an incorrect mobile phase (analyst error)

Decreases in pressure can be the result of

- a change in flow rate
- a small leak
- an incorrect column (analyst error)
- an incorrect mobile phase (analyst error)

## Diagnosis and Solutions

Because changes in pressure can be the result of an incorrect column, mobile phase, and/or flow rate, the first things to check are column specification, mobile phase composition, and flow rate. These are easy to check and might help to avoid costly instrument downtime. In the event that the pressure change is not the result of an error in column, mobile phase, or flow rate, the next course of action will depend on the nature of the problem.

**High Pressure**  In the event of high pressure, the system should be checked to identify a potential blockage. Particulate matter within the samples or small amounts of contamination in the mobile phase can cause blockages over a period of time. Particulate matter can build up in filters and on frits within the system. The most likely place for a blockage to occur is in the in-line filter, guard column, or the analytical column. Each of the components should be checked by loosening the fittings one by one, working upstream toward the pump. Any blocked components should be replaced or cleaned where necessary. It may be possible to back-flush the column in order to remove any particulate matter on the frit. In-line filters can be cleaned in a sonic bath using an appropriate solvent.

Precipitation of buffer salts can present a problem in situations where the organic portion of the mobile phase is greater than approximately 50%. This happens because the buffer salts are no longer soluble in the mixtures of organic solvents and water and 'salting-out' occurs. In these situations, the buffer portion of the mobile phase should be replaced with water and the system should be flushed until the pressure drops back to an acceptable level. The concentration and type of buffer should be reappraised through further method development.

**Low Pressure**    Low pressure is usually the result of a leak in the system. As mentioned previously, leaks normally occur around the column connectors or in the pump head. The column can easily be removed and the fittings tightened where necessary. The pump head should be inspected for signs of leaking mobile phase or salt deposits. Repairs can be made to the pump.

## Tool Kit and Spare Parts

A tool kit composed of spanners, screwdrivers, etc. designed for HPLC systems is a useful addition to your laboratory. This can usually be purchased from the HPLC manufacturer or may even be supplied with the system. It will contain all of the necessary tools to allow you to undertake routine preventative maintenance (discussed next). If a tool kit does not come with the system, most of the spanners and screwdrivers can be purchased from your local hardware store. It is recommended that you keep a tool kit handy.

In conjunction with the tool kit, it is prudent to maintain a supply of consumable items that can be used in an emergency or as part of a preventative maintenance programme. A list of common consumable items that you might expect to keep in the laboratory is given in Table 10.2.

## Preventative Maintenance

In the forensic environment, you might find yourself being asked to analyse a sample in relation to a large drug find or a sudden death. A suspect may have been incarcerated and your results are required quickly. When you are working to tight turnaround times, you need to be sure that your HPLC system works the first time without lengthy downtime due to breakdown of the system. The system therefore needs to be maintained as part

**Table 10.2   Recommended Consumable Items**

| Pump | Column | Injector | Detector |
|------|--------|----------|----------|
| Pump seals | Column frits | Rotor seal | Lamp |
| Pistons (sapphire) | | Syringe | Flow cell |
| Check valves | | Needle | |
| | | Needle seat | |
| | | Loop | |

**Table 10.3    Preventative Maintenance Programme**

| Part | Issues | Lifetime | Cleaning and Care |
|------|--------|----------|-------------------|
| Seals | Worn due to issues with piston Buffer salt deposits | Months | Replace the pump seals on a 3-month basis, depending on the nature of the mobile phase |
| Pistons | Scratches due to buffer salts | Years—with correct care | Inspect for scratches periodically and after changing the pump seals |
| Check valve | Contamination from mobile phase Contamination from pump seal Buffer salt deposits cause blockage | Years—with correct care Replace pump seals Flush with HPLC grade solvent | Use filters |

of a preventative maintenance programme. This programme significantly reduces downtime due to system failure.

You can never rule out sudden failure altogether. Changing some parts regularly greatly reduces the likelihood of sudden failure. Consumable parts are those that will wear or degrade naturally with time and use. An example would be pump seals; due to the constant back and forth movement of the pistons and to deposited buffer salts, these seals will eventually wear away and cause the pump to leak. Leaks in the system (discussed in greater detail earlier in the text) can cause variation in retention times due to fluctuations in the flow rate. In extreme cases, seal wear can cause the pistons to break, causing pump failure. It can be much more costly in the long run to wait until the system breaks down. Some might say, 'Don't fix it if it's not broken', but this is often a false economy in a high-pressure forensic environment. The exact nature of the preventative maintenance programme will depend on the laboratory, but Table 10.3 provides some general information in relation to pumps.

# Column Care

In order to enhance the column lifetime, it is essential that the columns are looked after and maintained.

## Storage

For short-term storage, such as overnight, it is appropriate to store the column in the mobile phase that has been used for the last assay. If the mobile phase contains a high proportion of buffer salt, the pump should be allowed to run at a reduced flow rate until the next use. For longer term storage—for example over a weekend—it is advisable to flush the system with a mobile phase containing water/methanol in the same proportions as the mobile phase used for the assay. This method is suitable for use when buffer salts have been present in the assay mobile phase. Introduction of the water to replace the buffer ensures that any salts are flushed from the system.

Flushing with pure methanol straight after buffer use will most likely result in the precipitation of buffer salts, which will cause the system to block and can be difficult to remove later. Once the buffer salts have been removed, it is possible then to flush the system with either methanol or acetonitrile for storage. If the column will not be used for a longer period of time, then buffer salts should be flushed out and the column stored in acetonitrile.

## Regeneration

In cases of a buildup of 'dirt' on the top of the column, the column can be treated as follows: Flush the column with a range of solvents with increasing elution power (starting with water and then acetonitrile, followed by isopropanol and heptane, back to isopropanol, and then acetonitrile). This will wash off any materials that may have stuck to the column due to nonpolar/nonpolar interactions. Another method is to turn the column around and back-flush with mobile phase. This acts to drive off any contamination from the top of the column.

---

### KEY POINT SUMMARY

**Quick Guide to HPLC Troubleshooting**

| Symptom | Cause | Solution |
|---|---|---|
| No signal | The detector is switched off | Switch on the detector |
|  | The lamp has blown | Install new lamp |
| No peak(s) | There has been an error in the sample preparation | Check sample preparation |
|  | There is a blockage in the injector | Check needle and clear any blockages |
|  |  | Check switching valve; call engineer if it is faulty |
|  |  | Check the syringe and clear any blockages |
|  | Vial error (misplaced) | Check vial position and cross-check with sample sequence |
|  | The pump is not switched on | Switch the pump on and ensure that mobile phase flows through system |

*(continued on next page)*

---

| Symptom | Cause | Solution |
| --- | --- | --- |
| Fluctuations in retention time | There is a leak | Perform systematic system check to identify source of leak |
| | | Detector: replace seals or gaskets |
| | | Injector: replace value rotor |
| | | Pump: replace piston seals, check piston for damage |
| | | Column: tighten fittings |
| | Mobile phase preparation error | Ensure that the mobile phase has been prepared correctly; cross-check with SOP. Prepare the mobile phase again if required |
| | Mobile phase proportioning valve mixing error | Prepare the mobile phase manually and rerun samples. Call engineer if required |
| | Evaporation of organic portion of mobile phase | Prepare mobile phase again, ensuring that the container is of an appropriate size and is covered |
| | The wrong column has been installed | Exchange the column and install the correct one |
| | The column is old and is no longer functioning as expected | Install a new column of the same type |
| | The system has not been allowed to equilibrate for a sufficient amount of time | Allow the mobile phase to flow through the column for at least 20 minutes |
| | The column is contaminated | Physically remove the contamination from the top of the column |
| | | Reverse the column and flush with mobile phase (ensure that the column outlet is disconnected from the system) |
| | | Remove the column and change the frit |
| | | Use a guard column |
| | There are fluctuations in room temperature | Maintain a steady temperature by installing a column heater |
| | | Ensure that ambient room temperature is held constant (air-conditioning) |
| Baseline noise | There is air in the system (flow cell) | Add a pressure restrictor to outlet tubing |
| | | Increase the mobile phase flow rate slightly to remove trapped air |

| Symptom | Cause | Solution |
|---|---|---|
| | Contamination | Remove the source of the contamination as described earlier |
| | | Change the mobile phase solvents; ensure that the correct grade is used |
| | The lamp has deteriorated | Replace the lamp |
| Poor resolution | Mobile phase preparation error | Ensure that the mobile phase has been prepared correctly; cross-check with SOP. Prepare the mobile phase again if required |
| | Mobile phase proportioning valve mixing error | Call engineer |
| | Evaporation of organic portion of mobile phase | Prepare mobile phase again, ensuring that the container is of an appropriate size and is sealed |
| | The wrong column has been installed | Check the column information and install the correct one |
| | The column is old and is no longer functioning as expected | Install a new column of the same type |
| | The system has not been allowed to equilibrate for a sufficient amount of time | Allow the mobile phase to flow through the column for at least 20 minutes |
| Peak tailing | Incorrect mobile phase pH | Prepare the mobile phase and adjust the pH to the correct value |
| | The column is old and is no longer functioning as expected | Install a new column of the same type |
| | There is a void at the top of the column | Repack the void with packing material or install a new column of the same type |
| | There is another peak partially coeluting with the peak of interest | Matrix effects: extract the sample using either LLL or SPE |
| | | Method: further development required to ensure adequate separation of peaks |
| Peak fronting | The sample is overloaded | Dilute the sample into the linear range of the method |
| | The sample solvent is incompatible with the mobile phase | Prepare the sample again in the correct solvent (the mobile phase should be used where possible) |

*(continued on next page)*

| Symptom | Cause | Solution |
|---|---|---|
| Broad peaks | Late elution of analytes carried over from previous injection | Method: further development required to ensure all peaks are eluted with given run time |
| | The injection volume is too large | Inject a smaller sample volume (sample preconcentration may be required) |
| Splitting peaks | There is a fault with the injector | Inspect the rotor assembly for damage and replace where necessary |
| | There is a void at the top of the column | Repack the void with packing material or install a new column of the same type |
| | There is contamination in the system (top of column) | Physically remove the contamination from the top of the column |
| | | Use a guard column |
| | | Reverse the column and flush with mobile phase (ensure that the column outlet is disconnected from the system) |
| | | Remove the column and change the frit |
| | The sample solvent is incompatible with the mobile phase | Prepare the sample again in the correct solvent (the mobile phase should be used where possible) |
| Increase in system pressure | There is dirt in the system, causing a partial blockage | Identify the source and location of the blockage. Remove the column and flush the system with water or a strong organic solvent |
| | The buffer salts have precipitated in the system | Identify the source and location of the blockage. Flush the system with mobile phase where no buffer salts have been added |
| | The wrong column has been used | Install the correct column |
| | The wrong mobile phase has been used | Use the correct mobile phase |
| | There has been a change in the flow rate | Check the method and ensure that the correct flow rate is used |
| Decrease in system pressure | There is a leak | Perform systematic system check to identify source of leak |
| | There has been a change in the flow rate | Check the method and ensure that the correct flow rate is used |

## Further Reading

Agilent Technologies. 2009. *LC chromatography troubleshooting: Frequently asked questions* (www.home.agilent.com).

Agilent Technologies. 2009. *LC column troubleshooting: Frequently asked questions* (www.home.agilent.com).

Dolan, J. W., and L. R. Snyder. 1989. *Troubleshooting LC systems: A comprehensive approach to troubleshooting LC equipment and separations.* Totowa, NJ: Humana Press.

Meyer, V. R. 2006. *Pitfalls and errors of HPLC in pictures.* New York: John Wiley & Sons.

Phenomenex. 2009. *HPLC troubleshooting guide* (www.phenomenex.com).

Sadek, P. C. 2000. *Troubleshooting HPLC systems—A bench manual.* New York: John Wiley & Sons, 2000.

Waters Corporation. 2002. *HPLC troubleshooting guide* (www.waters.com).

# Forensic Applications of HPLC

11

## Introduction

Forensic science encompasses a number of different fields of science. In this book, we are explaining the theories associated with high performance liquid chromatography, the different forms that it may take, and its use in some forensic applications. In this chapter, we will examine applications of HPLC in drug analysis, toxicology, analysis of explosives, analysis of coloured materials, and environmental science.

## Drug Analysis

### Introduction

Samples that may contain controlled drug substances can be submitted to the laboratory in a variety of forms. This includes trace samples and bulk samples. Examples of trace samples may include, but are not limited to, swabs from surfaces where drug substances may have been prepared or from paraphernalia that may have been used in the administration of a controlled substance (very small amounts of which may not be visible to the naked eye). Bulk samples may include, but are not limited to, tablets, powders, plant material, resins, and liquids.

### Sampling Techniques

Each of these different sample types will require different methods of preparation prior to analysis by HPLC. First, it is necessary to decide how much of the submitted sample should be analysed because there may be much of the same sample. For example, if 2,000 tablets all visually appear to be the same, should we analyse every single tablet? The answer here is no because we do not have the time to do all of the preparation and analysis of each tablet. Also, it is possible to take a smaller sample set that can be shown to be a statistically viable data set representative of the sample of tablets submitted to the laboratory for analysis.

Each lab can vary, sometimes quite considerably, but this is acceptable as long as they can provide evidence to support the fact that the size of the set is representative of the overall sample submitted. Guidelines for sampling are available from the United Nations Office on Drugs and Crime (UNODC).

The following sampling techniques are given in the UNODC guidelines and have been summarised here as an example. However, as previously mentioned, there is no legal requirement for a testing laboratory to stick to these techniques and many will use other statistically representative sampling methods.

The UNODC guidelines for sampling are based upon the statistical method known as hypergeometric distribution. Hypergeometric distribution is used to determine the probability of the number of positive samples likely to be found in a certain population. This is based on the following equation:

$$P\left(X=x\middle|N_1,N,n\right)=\frac{\dbinom{N_1}{x}\dbinom{N-N_1}{n-x}}{\dbinom{N}{n}}$$

where
  $P$ = probability
  $N$ and $N_1$ = in a population of $N$, $N_1$ is the number of positives
  $n$ = sample size
  $X$ = number of positives

Binomial and Bayesian statistical methods may also be used.

Generally, when we have packages, the following rules can apply (again, based on the UNODC guidelines):

| <10 packages: | all should be tested |
| >10 and <100: | test 10 chosen at random |
| >100 packages: | $\sqrt{\text{number of packages}}$ |

For tablets, the following rules can apply:

| 1–50 tablets: | half of the sample size (up to 20) |
| 51–100 tablets: | sample 20 |
| 101–1,000 tablets: | sample 30 |
| >1,000 tablets: | $\sqrt{\text{number of tablets}}$ |

For liquids of different phases, all phases should be tested.

**Table 11.1   Homogenisation Techniques Available in Drug Analysis**

| Homogenisation Technique | Method |
|---|---|
| Cone and square | Used with powdered samples. The materials are mixed, with any larger fragments reduced in size. From here, the material will be split into four sections on a clean, flat surface. Two of the 'squares' (opposite to each other) will be removed and the two left will be recombined with the procedure being repeated until an appropriate sample size for analysis is achieved. |
| Scraping | Used with tablet material. In this technique, tablets submitted will be scraped instead of completely powdered. This is done to keep the tablet intact for any further forensic analysis, whereas powdering would mean that the tablet would be lost. |
| Blending | This can be used with plant material and tablets. This technique involves the use of an electric blender to homogenise samples. The major drawback of using this technique is determining that the blender is clean before using it for another sample. |

## Homogenisation and Extraction Techniques

As well as sampling, another important aspect of sample preparation for drug analyses is the homogenisation of a subsequent extraction of the sample. This step is necessary, particularly if we have a powder that is not homogenous. A number of homogenisation techniques are available in drug analysis and these are summarised in Table 11.1.

## Analytical Techniques Employed

When a sample set has been determined to be statistically representative of the population submitted to the laboratory and homogenised when appropriate, the next step is to determine whether or not an extraction technique, such as liquid–liquid extraction (LLE) or solid phase extraction (SPE) is necessary (see Chapter 3 for explanation of these techniques). If the sample size is large enough and the type of drug substance that may be present is unknown, a series of presumptive tests can be carried out, as can a general unknown screen.

Presumptive tests are a series of colour tests that can provide an indication as to the class of drugs that might be present; however, because these tests are subjective, confirmatory analysis must also be carried out. A general unknown screen is commonly carried out using GC-MS, although LC-MS[n] has also recently found an application in screening.

These initial steps are carried out in order to determine the class of drugs and whether any drugs are present in the samples analysed. The main reason for this is that a different analytical method is required for different classes of drugs—whether gas chromatography or the liquid chromatographic technique is used. For example, if it was determined at the presumptive stage of testing that a sample may contain an opiate drug and benzodiazepine, it may be necessary for the sample to be quantified using two separate analytical methods: one for opiates and one for benzodiazepines. The reason for this is that the chemistry of the various drug substances present in the sample may be very different and to analyse the two classes of drugs using the same method may cause an extremely long run time, or it may not be possible to the levels of specificity, accuracy, and precision required.

## HPLC Analysis of Drug Samples

In February 2009, the British Broadcasting Corporation (BBC) reported that the Medicines and Healthcare Products Regulatory Agency (MHRA) recalled a number of medicines because counterfeit products had been introduced to the market that were found to contain between 50 and 80% of the required dose. The drugs recalled were Casodex®, which is used for the treatment of prostate cancer; Plavix®, which is used to treat heart conditions and strokes; and Zyprexa®, which is used to control symptoms of schizophrenia (chemical structures are shown in Figure 11.1).

The pharmaceutical industry complies with strict regulatory guidelines in order to ensure that pharmaceutical products released to patients are appropriate for administration. This means that all products will be tested before the batch can be released to the market; a series of efficacy tests are completed. When these tests are completed on the raw materials and the final preparation, certificates of analysis can also be completed.

The tests are carried out to pharmacopoeia standards (a pharmacopoeia is a technical text that contains tests for identification of products used in the pharmaceutical industry). For example, if we consider the previously mentioned products (chemical structures shown in Figure 11.1), they can be tested by HPLC-DAD under the European Pharmacopoeia (shortened to Ph Eur on packaging). Initially, the counterfeit products were identified by a wholesaler who noted a difference in the packaging; by comparing results of bona fide tablets and the suspected counterfeit products, differences in the dose can be identified (see Figures 11.2 and 11.3).

By examining the difference in the peak heights or by carrying out calibration and determining the actual dose present in the sample, problems with the batch can be identified.

Casodex®

Plavix®

Zyprexa®

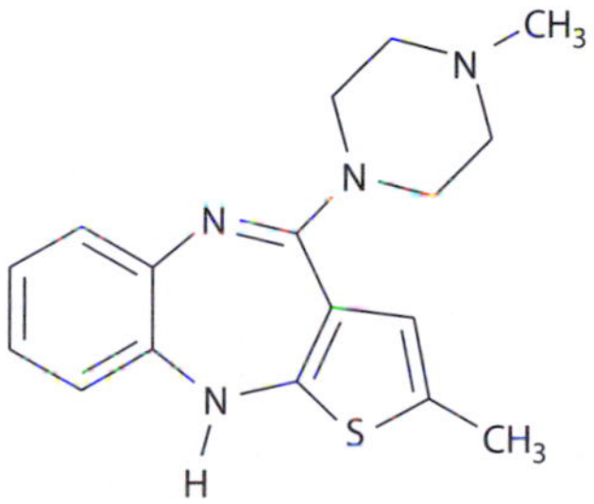

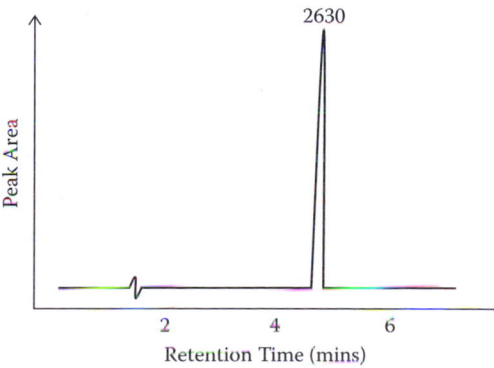

**Figure 11.1** Chemical structures for Casodex®, Plavix®, and Zyprexa®.

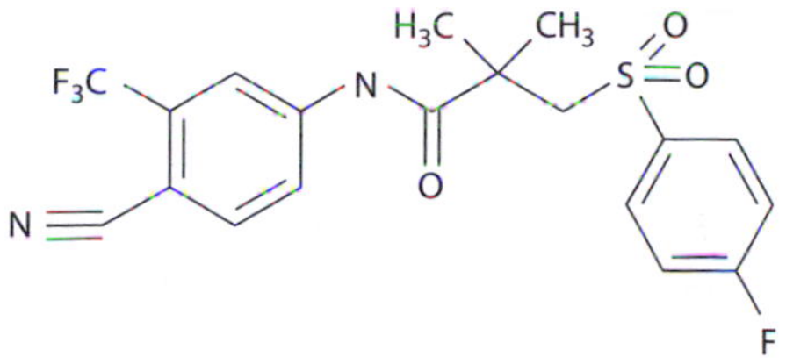

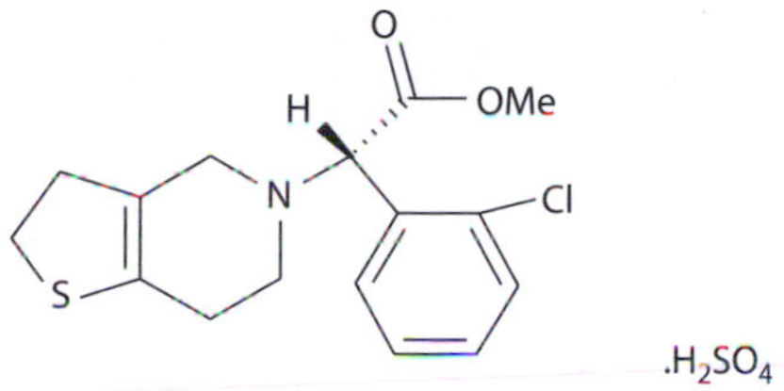

**Figure 11.2** Chromatogram of *bona fide* product.

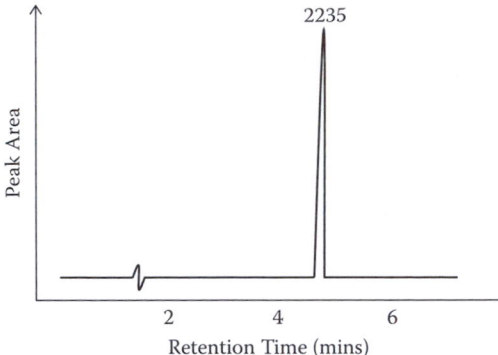

**Figure 11.3** Chromatogram of suspected counterfeit product.

## Toxicological Analysis

### Introduction

Forensic toxicology encompasses a number of different applications; these include alcohol and drugs in cases of suspected impaired driving, analysis of samples collected at the scene of a crime or from a postmortem examination, samples collected for workplace drug testing, sample detection of performance-enhancing drugs in athletes, and samples collected in cases of alleged drug-facilitated sexual assault. There are many differences between the samples and the manner in which they are collected in the these examples. However, the major difference here is whether or not the individual was alive or dead when the sample was collected because this can have a marked effect on the analyses carried out and the subsequent interpretation.

Many different samples may be submitted to the toxicology laboratory for analysis (see Table 11.2). When samples have been submitted from a

**Table 11.2  Samples That May Be Submitted for Analysis**

| Samples from the Deceased | Samples from Living Individual |
|---|---|
| Blood | Blood |
| Urine | Urine |
| Hair | Hair |
| Nails | Nails |
| Vitreous humour | Oral fluid |
| Tissue samples (such as liver, lung, brain) | |
| Gastric contents | |
| Bile | |

postmortem, the nature of the sample can be very different due to the state of decomposition in which the body was found. Decomposition can greatly affect the tests and interpretation that can be carried out on each of the samples for analysis (Karch 2003; Jones 2009; Drummer 2004, 2007).

## Sampling Techniques

When samples are taken from a living patient, there is no major issue because urine can be collected readily, blood samples can be taken by a trained phlebotomist, and hair (if this is possible to collect) and nails can be trimmed. However, sampling becomes more difficult when samples are collected postmortem. From the time that an individual dies, decomposition starts to take place. If a person dies in a hospital and is removed to the mortuary fairly quickly, refrigeration slows down the rate at which the body decomposes. However, if a person has been found outside or in a building and some time has passed before discovery of the body (days or longer), decomposition processes will not have been hindered.

When a person is alive, blood is continually pumped through and around the body by the heart; when a drug is administered, it will be absorbed into the bloodstream (the route by which this occurs ultimately depends upon the route of administration). The body processes this drug through a process known as ADME (absorption, distribution, metabolism, and elimination).

Consider, as an example, that a man ingests a tablet containing amphetamine; this drug will be absorbed through the wall of the stomach and into the flow of the bloodstream. From here, the blood distributes it to other parts of the body, such as the liver, where it is metabolised and subsequently turned into water-soluble or inactive metabolites that can be easily eliminated from the body in the urine. We know that the plasma half-life for amphetamine is between 4 and 8 hours; therefore, if we were to test a blood sample taken 2 hours after the initial administration, we would expect to find the parent drug (amphetamine) as well as associated metabolites.

When an individual dies, the blood is no longer pumped through the body, so it starts to 'pool' with gravity. Thus, if the person has died lying on his or her back, the blood will collect inside the body cavity at the back. Some drugs are prone to redistribution; this means that the drug will diffuse from reservoirs of high concentration in a solid organ, such as the liver, into the blood. This is why it is essential for a toxicologist to have as much information about where the sample was taken: Blood collected could have very different concentrations of drug substances, depending upon the sampling site. Table 11.3 shows the concentration of amitriptyline found in different samples collected from the same deceased person.

**Table 11.3    Concentrations of Amitriptyline Found in
Different Samples Collected from the Same Decedent**

| Sample | Amitriptyline Concentration (mg/L) |
| --- | --- |
| Femoral vein | 1.8 |
| Left and right iliac veins | 2.6 |
| Lower inferior vena cava 1 | 3.5 |
| Inferior vena cava | 2.4 |
| Superior vena cava | 2.8 |
| Right heart | 2.7 |
| Pulmonary artery | 3.4 |
| Left pulmonary vein | 18.8 |
| Right pulmonary vein | 20.2 |
| Aorta | 11.7 |
| Vitreous humour | 0.8 |
| Bile | 42.1 |
| Urine | 13.6 |

*Source:* Reproduced from Langford, A. M., and D. J. Pounder.
1997. *Journal of Forensic Science* 42 (1): 88–92.

## Analytical Techniques Employed

The main steps involved in toxicological analyses are to detect, identify, quantify, and subsequently interpret the analytical findings based upon the information provided. Generally, the detection is carried out as a screening step in order to establish whether any group of substances may be present. These tests typically include immunoassay techniques such as enzyme multiplied immunoassay technique (EMIT) or enzyme linked immunosorbent assay (ELISA). These techniques are based on antibody–antigen interactions. Colour tests and thin layer chromatography may also be used at this stage. GC-MS may also be used in cases when no information is available and a general unknown screen is required to pinpoint substances.

As previously mentioned, these tests will provide information on which types of substances may be present, but they do not provide specific and sensitive results for particular substances. For example, a urine sample that is tested using ELISA indicates that opiates may be present in the sample; however, at this point, it is not possible to differentiate morphine, codeine, or diacetylmorphine (active component of heroin). Further analysis is required to determine this.

The next step is identification and quantitation (if required), in which a group of substances, such as opiates, has been established in the initial screen. Analytical techniques commonly used include GC-MS, LC-MS, and HPLC-UV. These techniques are used in quantitation as well as detection.

## HPLC in Toxicological Analyses Examples

Until recently, GC-MS was the 'gold standard' technique used in identification and quantitation in toxicology; however, LC-MS has now 'found its feet', so to speak, and is finding more applications in this field. As a technique, LC has many advantages over GC—one of which is that derivatisation (a sample preparation technique carried out in order to produce a thermally stable compound suitable for analyses at high temperatures used in gas chromatography) is not required because LC analyses are carried out at room temperature. This is an advantage because the process is time consuming and will be carried out as well as homogenisation and extraction techniques; this adds to the time taken to analyse one sample.

In January 2008, Australian actor Heath Ledger was found dead, and toxicology analysis revealed the presence of prescription medication. Oxycodone and hydrocodone (opioid analgesics); diazepam, temazepam, and alprazolam (benzodiazepines); and doxylamine (antihistamine) were all found in the samples from the deceased. The coroner in the case concluded that death was due to the combined effects of these drugs. As mentioned, GC-MS has been the gold standard for these types of analyses; however, LC-MS/MS in particular is being reported in the scientific literature.

Restek, a manufacturer of chromatographic columns and other chromatography supplies, has provided methodology for the analysis of opiates by LC-MS/MS, as have a number of scientists in academic journals (Sellers 2007; Coles et al. 2007; Musshoff et al. 2006). The detection of benzodiazepines, including those mentioned previously, has also been reported; a technical note was recently released from Agilent Technologies providing methodology for the analysis of benzodiazepines in oral fluid (Moore et al. 2007).

The use of hyphenated LC systems has taken off over the last few years. As a profession, we still have some way to go before LC-MS and LC-MS$^n$ find their place in analysis as GC-MS has done.

## Colour Analysis

### Introduction

Fibres are ubiquitous; they are found in the clothing that we wear and the upholstery of our homes, places of work, and vehicles. Fibres can also be transferred from surfaces or from people, depending upon how much the fibres from the garment being worn shed and the ability of the receiving garment to accept those fibres (transfer and persistence of fibres).

Fibres are the single strands of a polymer, whether natural or synthetic, that are spun together to produce yarns used in the manufacture of fabric to produce garments and upholstery (as well as other applications). These fibres or yarns will have been treated before they are used to produce the fabric, and part of this treatment will involve the application of colour or dyeing in order to make the garments and fabrics more appealing to the consumer. Due to their commonality, it is not surprising that fibres are frequently submitted to the forensic laboratory for analysis.

## Sample Types

Fibres may be submitted in a tape lift format, which is a piece of sticky adhesive tape called J-Lar®. This tape is less adhesive than cellotape and is used to collect extraneous fibres from clothing or from the upholstery inside vehicles, for example. Any fibres that are different from the background fibres can be removed later for subsequent examination and comparison with fibre from other items involved in a case. Items of clothing and hair combings are also frequently submitted for examination for the presence of fibres.

## Analytical Techniques Employed

Most of the techniques employed in fibre analysis are nondestructive tests to determine whether the fibre is natural (obtained from animal, plant, or mineral) or synthetic (wholly manufactured from chemicals or regenerated from natural fibres) and the fibre type (e.g., determining if the fibre is wool, cotton, nylon, polyester, etc.). Whether any chemical treatments have been carried out (such as bleaching or the use of delustrants) is noted and the colour is also determined. Many of the techniques commonly used in these analyses include low- and high-power microscopes, Fourier transform infrared (FTIR) microscopy, polarising light microscopy, fluorescence microscopy, and microspectrophotometry (MSP).

Low- and high-powered microscopes are used to examine the morphological features of the fibres and the initial determination of whether the fibre is natural or man-made. FTIR microscopy can be used on a synthetic fibre to provide information in relation to the functional groups present; this can be used to pinpoint which synthetic fibre it is. Polarising light microscopy is used with synthetic fibres; plane-polarised light interacts with the fibres in order to provide refractive index values (many of these fibres have two refractive indices due to the chemical structure of the fibre and are said to be birefringent). This helps in the identification of the synthetic fibre.

Fluorescence microscopy can be used to determine whether delustrants and other similar chemical treatments have taken place on the fibre, and microspectrophotometry is used to determine the chromaticity coordinates for the colour of the fibre. This latter technique is good for determining the chromaticity coordinates of fibres; however, it does have some limitations with respect to deep colours.

## HPLC in Colour Analyses

The chromaticity coordinates cannot provide information in relation to the ratio of coloured dyes present in the fibre, so this is where techniques such as thin layer chromatography (TLC) and HPLC can be used. It should be noted, however, that these techniques are used only when the results can be used to assist in an investigation; they are not used routinely because, in order to prepare the fibre for analysis, the dye would have to be extracted from it and the process would therefore be deemed destructive. For TLC, a reasonable amount of dye would also be required; this means that, on small sample sizes where only one or two fibres are available, this technique is not a viable option.

# Analysis of Explosives

## Introduction

An explosion occurs when energy that was previously confined is suddenly released to affect the surroundings. Small-scale explosions, such as a shaken can of fizzy juice exploding, are harmless; however, explosions such as those produced to demolish a building are an example of the opposite extreme of a scale. Explosives are chemical compounds or mixtures of chemical compounds in which much energy is stored; this energy should be able to be released quickly in order to provide the energy to be classed as an explosive.

An explosive can be defined as 'a sudden or violent release of physical or chemical energy, often accompanied by the emission of heat, light, and sound' (White 2004). Explosions can be characterised by their source of energy (i.e., whether that source is physical or chemical) and also by whether the explosion is in the dispersed or condensed phase. The following are examples of the different types of classes of explosion:

- physical: an exploding pressure vessel (e.g., a hairspray canister in a fire)
- chemical: an explosion of a mass of sodium chlorate
- dispersed: detonation of a cloud of flour in the air in a bakery
- condensed: detonation of trinitrotoluene (TNT)

Explosives can also be classified by the detonation velocity at which they release the confined energy. This classification differentiates explosives as

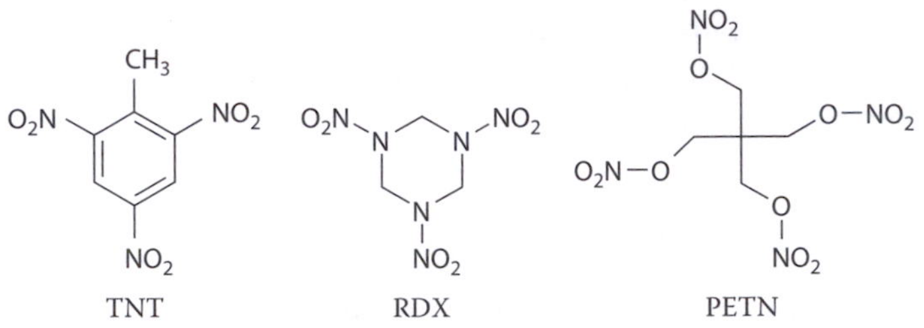

**Figure 11.4** Chemical structures of TNT, RDX, and PETN.

either *low* (deflagrating) or *high* (detonating). The difference between the two is that low explosives have a detonating velocity of $1 \times 10^{-3}$ of a second, whereas with high explosives, the detonating velocity is as much as 1,000 times faster than this ($1 \times 10^{-6}$ of a second).

The detonating velocity and therefore the amount of energy released quickly will limit the applications of each type of explosive. Low explosives are typically used as propellant I firearms ammunition or in pyrotechniques (fireworks). High explosives, on the other hand, are typically used when large amounts of destruction are required, such as in military applications or in the demolition of buildings. Examples of high explosives include tri-nitrotoluene (TNT), cyclotrimethylenetrinitramine (RDX), and pentaeryth-ritol tetranitrate (PETN), the chemical structures of which can be seen in Figure 11.4.

## Sample Types

The types of samples that are submitted to the forensic science laboratory for examination for the presence of explosives can include samples collected from the scene of the incident, from suspects, and from any sites where explosives are suspected to have been manufactured or processed. These samples can include debris collected from the site of the incident, remains of a detonating device, and clothing and hand swabs from suspects. (Other samples may also be collected, such as raw material that could have been used in the manufacture of explosives or explosive devices.)

## Techniques Employed

Many techniques can be employed in the analysis of compounds that can be used in the manufacture of explosives and explosive materials. Raman and infrared spectroscopy, nuclear magnetic resonance (NMR), mass spectrometry, and gas chromatography can all be used in the analysis of this type of material. Hyphenated chromatographic techniques, such as GC-MS and LC-MS, can also be used.

## HPLC in Explosive Analyses Examples

In July 2005, London was subjected to a series of terrorist attacks: Bombs exploded on the public transport network at rush hour. Three bombs were detonated on trains in the London Underground and one was detonated on the upper deck of a bus. It was initially thought that military-grade plastic explosives had been used in the attack on 7 July because it appeared that the explosions had been synchronised. However, it was later determined that the bombs were homemade devices consisting of triacetone triperoxide (TATP). TATP is a peroxide-based explosive that is highly susceptible to shock, heat, and friction and is one of if not the most sensitive explosives known (Bubnikova et al. 2005).

Peroxide-based explosives, such as TATP, diacetone diperoxide (DADP), and hexamethylene triperoxide diamine (MHTD) can be detected using HPLC-DAD at 214 nm (see Figures 11.5 and 11.6). Successful LC-MS/MS of explosives has been reported, as has the use of ion mobility and liquid chromatography with amperometric detection (Vigneau and Machuron-Mandard 2009; Hilmi et al. 1999; Ou et al. 2009; Meng et al. 2008).

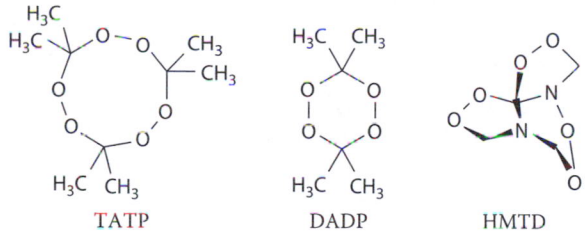

**Figure 11.5** Chemical structures of TATP, DADP, and HMTD.

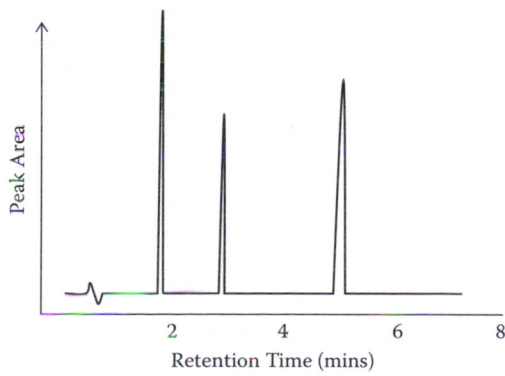

**Figure 11.6** Peak 1: DADP; Peak 2: TATP; Peak 3: HMTD.

## Food and the Environment

Food science and environmental science may not jump out immediately as being of forensic interest. An increase in legislation designed to protect both the environment and the consumer has meant that there have been many more prosecutions of companies and individuals who are thought to have breached regulations. Instances where there has been an issue with the use of illegal food additives or where illegal waste or effluent has been identified need to be investigated and a cause assigned. A number of high-profile incidents of this nature have been highlighted in the media, so let us look at some examples and consider each on its own merits in terms of HPLC analysis.

### Sudan Red Dyes

Sudan red dyes are synthetically produced dyes normally used for colouring shoe and floor polish, oils, and solvents. For example, Sudan red is used to colour petroleum products to prevent misuse. The dye is added to diesel products to prevent substitution of less expensive fuel oils. Some of the Sudan dyes have been used in the past to colour foodstuffs; however, many of these, including Sudan I (see Figure 11.7), Sudan III, and Sudan IV are now banned substances. Sudan dyes have been shown to cause cancer in animal studies and thus have been declared carcinogens. They are not permitted as an additive in any foodstuff in the United Kingdom and the rest of the European Union.

However, because something is a banned substance does not stop some people from using it illegally. It became apparent that some of the dyes had been used illegally to enhance the colour of chilli powder. A contaminated batch was imported from India and was used in the manufacturing process for Worcester sauce, which in turn had been added as an ingredient to many other foodstuffs. This led to a number of food recalls; as a result, any chilli or hot chilli foodstuffs imported into the EU must be accompanied by a certificate of analysis declaring the raw material or product free from Sudan red. Thus, foodstuffs will require examination in order to ensure compliance.

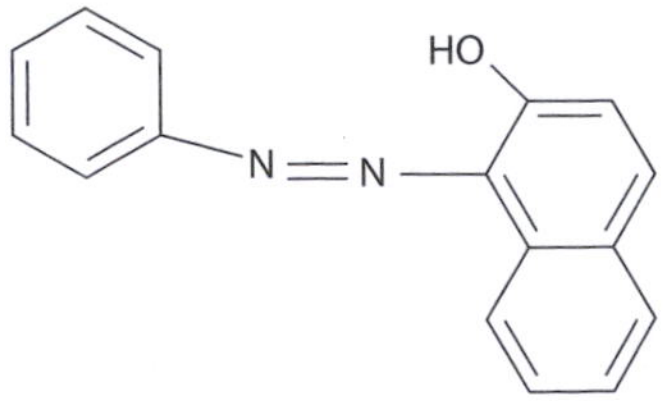

Sudan 1

**Figure 11.7** Chemical structure of Sudan I.

If we are to examine any foodstuff using HPLC, we will need to remove the matrix prior to analysis. The Sudan red dyes will need to be extracted from the matrix. Food matter can be a complex mixture of animal and/or vegetable material containing a cocktail of many different compounds. Oil products contain complex mixtures of nonpolar hydrocarbon materials that are incompatible with most HPLC modes on their own. Thus, extraction of each matrix will be required before any analysis of the Sudan red dye is undertaken by HPLC.

Solid phase extraction (see Chapter 3) has been successfully used to extract Sudan dyes from spices using a surface-modified polymeric-based column that has hydrophilic, hydrophobic, or π–π retention mechanisms. This particular phase can be used to extract both polar and nonpolar compounds. An example of a Sudan red dye is given in Figure 11.7.

These are nonpolar or hydrophobic molecules and the large number of chromophores present in the molecules means that there will be strong absorbance within the UV-visible (UV/Vis) range of the electromagnetic spectrum. This will facilitate the quantitative assessment using a diode array UV/Vis detector. Sudan dyes have been successfully analysed using $C_{18}$ packing materials, but because of the hydrophobic nature of the compounds, a high proportion of organic modifier is required in the mobile phase.

## Melamine in Baby Milk

Melamine (see Figure 11.8) is a nonpolar, weakly basic compound with a pKa of 5; it is commonly used as a fire retardant in polymer resins. It is toxic in nature and chronic exposure is thought to cause cancer or reproductive damage. In 2008, the BBC reported that many Chinese babies had become ill after drinking contaminated milk products. The contaminant was found to be melamine, which had been added to the baby milk to make it appear to have a higher protein content. Many batches of the contaminated milk have now been withdrawn from sale, but many babies have suffered as a result. When dairy products are examined by HPLC, it is necessary first to remove any interfering matrix-related substances. Sample preparation might include the following steps:

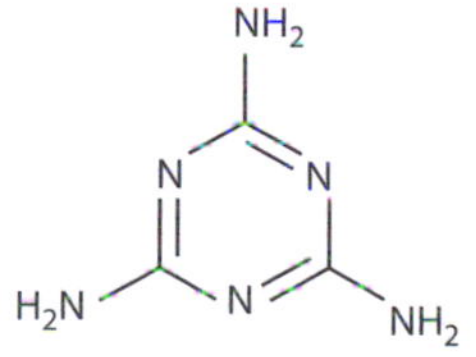

**Figure 11.8** Chemical structure of melamine.

- Remove the fat and protein by precipitation.
- Perform SPE or LLE to extract the melamine.
- Evaporate the extract to dryness.
- Reconstitute the residue in mobile phase or compatible solvent.

Melamine has been successfully analysed in baby milk using an acetonitrile precipitation followed by a filtration step. Other methods employed precipitation with trichloroacetic acid and acetonitrile followed by an SPE cleanup step using a strong cation exchange (SCX) cartridge. $C_{18}$ columns have been used in the HPLC phase of the analysis, with mobile phases buffered at pH 3 and containing ion-pairing reagent. We would expect that melamine would be completely ionised at this pH. This means that melamine will have very little, if any, retention on a $C_{18}$ column. To counter this, the ion-pairing agent is added and acts to make the compound more nonpolar and thus increase the retention time.

## Polyaromatic Hydrocarbons

Polyaromatic hydrocarbons (PAHs) are produced during the pryrolysis and incomplete burning of coal, oil, gas, and other organic materials. They are produced naturally in small amounts during forest fires and volcanic eruptions, but are considered to be man-made contaminants affecting the natural environment. Polyaromatic hydrocarbons are made up of fused aromatic rings with five- or six-membered rings the most common. They absorb in the UV region of the electromagnetic spectrum due to the large number of double bonds present in the ring structures (see Figure 11.9). They can be washed from the ground into the sea and are considered a serious pollutant and problem for marine life. They are thought to be mutagenic, to cause tumours, and to cause reproductive damage. Their release into the atmosphere is restricted, so monitoring is extremely important.

Sample cleanup or matrix removal is required before samples can be analysed by HPLC. The key feature of PAH compounds is that they are nonpolar aromatics. The more fused rings that are present, the more nonpolar

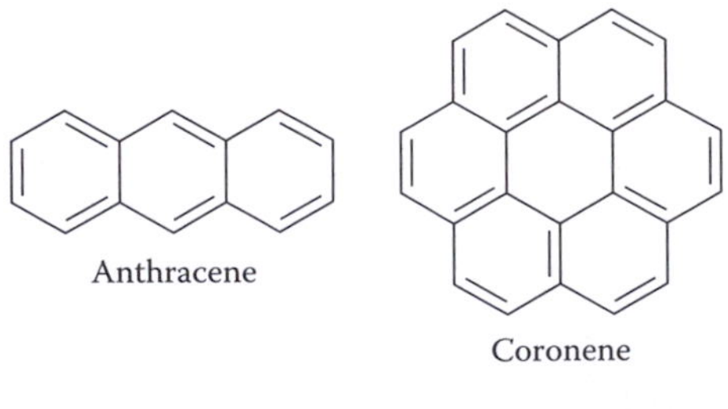

Anthracene

Coronene

**Figure 11.9** Chemical structures of anthracene and coronene.

the compound is. The PAHs in soil samples have been successfully extracted using successive amounts of toluene and ultrasonication. Toluene is used here due to the 'like dissolves like' principle because toluene is a nonpolar solvent. The PAHs can then be isolated using a silica-based fractionation column prior to analysis by HPLC. The chromatographic process is performed on a $C_{18}$ column using a combination of acetonitrile and water as the mobile phase.

The PAHs will have a high affinity for a nonpolar stationary phase such as the $C_{18}$ column used in the example. The greater the number of fused rings, the more nonpolar is the compound and the longer is the retention time. Separation will occur based on the number of rings present. To combat excessive retention times associated with the larger PAHs, a gradient elution method is used with increasing levels of acetonitrile in the mobile phase. Gradient elution involves increasing the amount of organic modifier as the run proceeds. This allows very nonpolar solutes to be eluted within a reasonable time frame without compressing the earlier part of the chromatogram and thus compromising resolution. Detection of the compounds is carried out using a UV/Vis detector at 254 nm.

## Pesticides

Pesticides are used to control cockroaches and other undesirable pests. They come in many forms, such as ant powders, insect repellents, and slug pellets. They are used in the environment and in the household to ensure that our food supplies are protected while they are growing and while they are stored in our fruit bowls and vegetable racks. Pesticides are effective pest killers, but are also harmful to humans; therefore, the use of these materials is covered by legislation and only those permitted within a country can be used legitimately at a controlled level.

There are those who breach the legislation and sell or use illegal pesticides at prohibited levels. Illegal pesticides often contain excessive amounts of known pesticides or, in some cases, unknown materials. They claim to be harmless to humans when in fact the opposite can be true. In some reported cases, pesticides have been found in supermarket produce in excess of the permitted levels. For example, the growth hormone chlormequat was found in pears and excessive levels of the organophosphate malathion were found in lettuce (see Figure 11.10).

Due to the ionic nature of chlormequat, ion exchange chromatography using an SCX column has been used successfully to analyse this compound. No chromophores are present in the molecule; therefore, a UV/Vis detector would not be suitable. Derivatising agents have been used to form colourimetric complexes, which can be measured using visible spectrophotometry. Mass spectrometry has also been used successfully to determine this

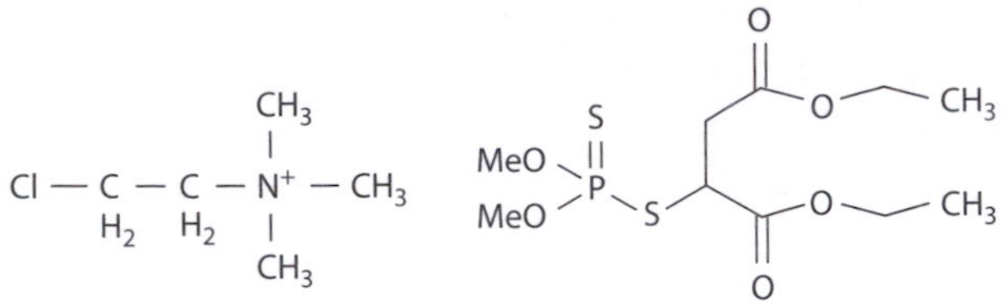

$$Cl - \underset{H_2}{C} - \underset{H_2}{C} - \underset{\underset{CH_3}{|}}{\overset{\overset{CH_3}{|}}{N^+}} - CH_3$$

Chloromequat

Malathion

**Figure 11.10** Chemical structures of chlormequat and malathion.

compound. Organophosphates such as malathion have been successfully analysed using $C_{18}$ columns with mixtures of acetonitrile and buffer (pH 3) in the mobile phase.

# References

Bubnikova, F., R. Kosloff, J. Almog, Y. Zeiri, R. Boese, H. Itzhaky, A. Alt, and E. Keinan. 2005. Decomposition of triacetate triperoxide is an entropic explosion. *Journal of the American Chemical Society* 127 (4): 1140–1147.

Coles, R., M. M. Kushnir, G. J. Nelson, A. Gwendolyn, and F. M. Urry. 2007. Simultaneous determination of codeine, morphine, hydrocodone hydromorphone, oxycodone, and 6-acetylmorphine in urine, serum, plasma, whole blood, and meconium by LC-MS/MS. *Journal of Analytical Toxicology* 31 (1): 1–14.

Drummer, O. H. 2004. Postmortem toxicology of abused drugs. *Forensic Science International* 142 (2–3): 101–113.

Drummer, O. H. 2007. Postmortem toxicology. *Forensic Science International* 165 (2–3): 199–203.

Hilmi, A., J. H. T. Luong, and A-L. Nguyen. 1999. Determination of explosives in soil and ground water by liquid chromatography-amperometric detection. *Journal of Chromatography A* 844:97–110.

Jones, A. W. 2009. In press. Postmortem toxicology is not quackery when done by qualified practitioners. *Journal of Forensic Science and Legal Medicine*.

Karch, S. B. 2003. Is postmortem toxicology quackery? *Journal of Clinical and Forensic Medicine* 10 (3): 197–198.

Langford, A. M., and D. J. Pounder. 1997. Possible markers for postmortem redistribution. *Journal of Forensic Sciences* 41 (7): 88–92.

Meng, H-B., T-R. Wang, B-Y. Guo, Y. Hashi, C-X. Guo, and J-M. Lin. 2008. Simultaneous determination of inorganic anions and cations in explosive residues by ion-chromatography. *Talanta* 76:241–245.

Moore, C., C. Coulter, K. Crompton, and M. Zumwalt. 2007. Determination of benzodiazepines in oral fluid by LC-MS/MS. Application note from Agilent Technologies (www.chem.agilent.com/Library/applications/5989-7201EN.pdf).

Musshoff, F., J. Trafkowski, U. Kuepper, and B. Madea. 2006. An automated and fully validated LC-MS/MS procedure for the simultaneous determination of 11 opioids used in palliative care, with 5 of their metabolites. *Journal of Mass Spectrometry* 41 (5): 633–640.

Ou, X.-K., X.-H. Yang, Y.-Q. Chen, and M.-C. Jin. 2009. Characterisation and deter-
    mination of chlorophacinone in plasma by ion chromatography coupled with
    ion rap electrospray ionisation mass spectrometry. *Biomedical Chromatography*
    23:524–530.
Sellers, K. 2007. Simplify and speed up opiate analysis. Restek publication (www.
    restek.com/advantage/adv_2007_04-05.pdf).
Vigneau, O., and X. Machuron-Mandard. 2009. An LC-MS method allowing the
    analysis of HMX and RDX present at the pictogram level in natural aqueous
    samples without a concentration step. *Talanta* 77:1609–1613.

## Further Reading

### Drugs

Cole, M. D. 2003. *The analysis of controlled substances.* New York: John Wiley & Sons.
Smith, F., and J. A .Siegel, eds. 2004. *Handbook of forensic drug analysis.* San Diego:
    Academic Press.

### Toxicology

Flanagan, R. J., A. A. Taylor, I. D. Watson, and R. Whelpton. 2008. *Fundamentals of
    analytical toxicology.* New York: Wiley-Blackwell.
Jickells, S., and A. Negrusz, eds. 2008. *Clarke's analytical forensic toxicology.* London:
    Pharmaceutical Press.
Karch, S. B., ed. 2007. *Postmortem toxicology of abused drugs.* Boca Raton, FL:
    CRC Press.
Moffat, A. C., M. D. Osselton, B. Widdop, and L. Y. Galichet, eds. 2003. *Clarke's analy-
    sis of drugs and poisons,* 3rd ed. London: Pharmaceutical Press.

### Explosives

Akhavan, J. 2004. *Chemistry of explosives,* 2nd ed. Cambridge, England: Royal Society
    of Chemistry.
Gaurav, V., A. Kaur, A. Kumar, A. Kumar Malik, and P. K. Rai. 2007. SPME-HPLC: A
    new approach to the analysis of explosives. *Journal of Hazardous Materials* 147:
    691–697.

### Colour

Alves, S. P., D. M. Brum, E. C. B. de Andrade, and A. D. P. Netto. 2008. Determination
    of synthetic dyes in selected foodstuffs by high performance liquid chromatog-
    raphy with UV-DAD detection. *Food Chemistry* 107: 489–496.
Christie, R. M., R. R. Mather, and R. W. Wardman. 1999. *The chemistry of colour appli-
    cation.* Chichester, England: Wiley Blackwell.
Gennaro, M. C., C. Abrigo, and G. Cipolla. 1994. High performance liquid chro-
    matography of food colours and its relevance in forensic chemistry. *Journal of
    Chromatography A* 674: 281–299.

Grieve, M., and J. R. Robertson, eds. 1999. *Forensic examination of fibres*, 2nd ed. New York: Taylor & Francis.

Zlotnick, J. A., and F. P. Smith. 1999. Chromatographic and electrophoretic approaches in ink analysis. *Journal of Chromatography B* 733: 265–272.

## Food and the Environment

Cardoso, A. S., S. A. Feliciano, M. H. Rebelo, S. S. Jose, and C. Reis. 2008. Optimisation and validation of a chromatographic methodology for the quantification of PAHs in drinking water samples. *WIT Transactions on Ecology and the Environment* 11: 271–272.

Delgado, B., V. Pino, J. H. Ayala, V. González, and A. M. Afonso. 2004. Non-ionic surfactant mixtures: A new loud-point extraction approach for the determination of PAHs in seawater using HPLC with fluorimetric detection. *Analytica Chimica Acta* 518 (1–2): 165–172.

Gratzfeld-Huesgen, A. *Analysis of Sudan red in diesel oil using HPLC*. Agilent Technologies (online).

Hou, S., D. Hwang, and H. Lee. 2003. High-performance liquid chromatographic determination of cyromazine and its derivative melamine in poultry meats and eggs. *Journal of Food and Drug Analysis* 11 (4): 290–295.

http://www.dionex.com/en-us/webdocs/70949_AN224-HPLC-Melamine-MilkPwd-18Mar09-LPN2184.pdf

http://www.govtlab.gov.hk/g/texchange/melamine.pdf

http://www.wwf.org.uk/filelibrary/pdf/mu_32.pdf

Kowalski, J. *Simple HPLC analysis for Sudan dyes monitoring Sudan I, II, III, and IV in a single, isocratic analysis*. Restek Chromatography Products. Available online.

Lautié, J. P., V. Stankovic, and G. Sinoquet. 2000. Determination of chlormequat in pears by high-performance thin layer chromatography and high-performance liquid chromatography with conductimetric detection. *Analusis* 28: 155–158.

Mudge, S. M. 2008. Environmental forensic and the importance of source identification. *Issues in Environmental Science and Technology* 26, Royal Society of Chemistry.

Nolett, L. M. L., ed. 2000. *Food analysis by HPLC,* 2nd ed. Boca Raton, FL: CRC Press.

*Polynuclear aromatic hydrocarbons by HPLC*. Available at http://www.cdc.gov

Ray, S., P. S. Khillare, T. Agarwal, and V. Shridhar. 2008. Assessment of PAHs in soil around the International Airport in Delhi, India. *Journal of Hazardous Materials* 156 (1–3): 9–16.

*Sample preparation and HPLC analysis of PAHs in extracted soil samples*. Available at http://www.knauer.net

Startin, J. R., S. J. Hird, M. D. Sykes, J. C. Taylor, and A. R. C. Hill. 1999. Determination of residues of the plant growth regulator chlormequat in pears by ion-exchange high performance liquid chromatography-electrospray mass spectrometry. *Analyst* 124: 1011–1015.

Sugita, T., H. Ishiwata, K. Yoshihira, and A. Maekawa. 1990. Determination of melamine and three hydrolytic products by liquid chromatography. *Bulletin of Environmental Contamination and Toxicology* 44 (4): 567–571.

## General

Langford, A., J. Dean, R. Reed, D. Holmes, J. Weyers, and A. Jones. 2010. *Practical skills in forensic science,* 2nd ed. Englewood Cliffs, NJ: Prentice Hall.

White, P. C., ed. 2004. *Crime scene to court: The essentials of forensic science,* 2nd ed. Cambridge, England: Royal Society of Chemistry.

# Glossary

**Accuracy:** A measure of the degree of closeness of the measured value to the true or actual value

**Activity coefficient:** A thermodynamic factor used to account for deviations from the ideal behaviour in a mixture of substances

**Adsorption chromatography:** Involves the interactions of a solute at the surface (or on fixed sites) of a solid stationary phase

**Amphiprotic:** A description given to a substance that can act as an acid and as a base (e.g., water)

**Analyte:** Substance or compound of interest measured in an analytical procedure

**Anion:** An ion or group of ions carrying a negative charge

**Atom:** Basic unit of matter with a central nucleus and surrounded by a cloud of negatively charged electrons

**Aufbau principle:** This principle describes the situation where atomic orbitals are filled with electrons one by one, starting with the orbital of lowest energy

**Buffer:** A solution that resists changes in pH when small amounts of acid or alkali are added to it

**Calibration:** The comparison of one measurement of known amount made on a specific piece of instrumentation with a second measurement made on a similar piece of equipment

**Capacity factor (retention factor):** A measure of the time the analyte resides in the stationary phase relative to the time it resides in the mobile phase

**Carryover:** That which is carried over or extended to a later time. In chromatography, this refers to material that is carried over from one run to another as a result of an insufficiently long run time or through contamination of the injector

**Cation:** An ion or group of ions carrying a positive charge

**Chiral:** A molecule that is not superimposable on its mirror image

**Chromatogram:** The pictorial representation of separated substances obtained using chromatography

**Chromatograph:** A piece of equipment used to generate a chromatogram or the act of separating a mixture of compounds using chromatography

**Column:** The solid support in which a chromatographic separation occurs

**Degassing:** The process of removing dissolved air under vacuum from a liquid, usually the mobile phase in HPLC

**Dilution:** Reduction in concentration of a solution through the addition of further solvent, usually to a known final volume

**Dipole–dipole moment:** Inter- or intramolecular interaction of molecules or groups having a permanent electric dipole moment

**Dipole moment:** Measured polarity of a polar covalent bond

**Dissociation:** The process by which a chemical combination splits up into its chemical components

**Dissociation constant:** A measure of the likelihood of a larger entity breaking up into smaller components. It is denoted by $K_d$ and the higher the value is, the higher the proportion of the dissociated material will be present in a mixture

**Distribution coefficient:** A measure of the distribution of an analyte between two phases. It is calculated as a ratio of the concentration of the analyte in one phase to the concentration in the second phase. In HPLC, the phases are a solid (stationary) phase and a liquid (mobile) phase

**Draw speed:** The speed at which the syringe on an autosampler draws the sample into the barrel. It can be adjusted to accommodate viscous samples

**Eddy diffusion:** The process by which substances are mixed due to eddies, where an eddy is described as being a current that is inconsistent with the main stream in a flow of liquid or gas

**Electron cloud:** The area around the atomic nucleus where the electrons are thought to reside

**Electron configuration:** The arrangement of electrons in an atom or molecule

**Element:** A pure chemical substance consisting of atoms that have the same atomic number

**Exclusion chromatography:** Relies on the ability of a porous solid stationary phase to discriminate on the basis of size by admitting small molecules to its pores but excluding larger ones

**Extraction:** The process of separating a substance from a mixture of substances

**Filtration:** A technique used to remove impurities from a solution or to isolate a particular chemical substance from a solution based on size

**Functional group:** A specific group of atoms within a molecule that characterise it in terms of reactivity. An example of a functional group is a carboxylic acid group (–COOH)

**Heisenberg uncertainty principle:** This principle states that 'it is impossible to specify simultaneously, with arbitrary position, the exact position and momentum of a particle'

**Hydrophilic:** Compounds that are usually charged or polar in nature and have an affinity for water

**Hydrophobic:** Hydrophobic compounds are usually neutral in nature and are repelled by water

**Ion exchange chromatography:** The process of separating compounds (ions or charged polar molecules) based on the charge–charge interactions between the sample and charges immobilised on the ion exchange resin

**Instantaneous dipole moment:** This type of dipole occurs when electrons are more heavily concentrated in one region of the molecule, thus creating a temporary dipole. This occurs in nonpolar molecules

**Intermediate precision:** Expresses the variation in results within a laboratory due to differences in (a) the instrumentation used, and (b) the analyst who carries out the processes

**Intermolecular forces:** Momentary unstable forces that act between stable molecules or between functional groups of macromolecules

**Intramolecular forces:** Any force that holds atoms or ions together in a molecule or compound. They can be covalent, ionic, or metallic

**Linearity:** A linear relationship in HPLC is demonstrated when the plot of the detector response as a function of concentration or content is found to be a straight line by statistical means. The linearity of an analytical procedure is its ability to obtain results directly proportional to the concentration of analyte in the substance (ICH Q2 R1)

**Lipophilic:** A substance that has an affinity for lipids; that is, it will dissolve much more readily in lipids (oily organic compounds) than it will in water

**Liquid–liquid extraction:** The process of separation of an analyte or analytes from a substance due to unequal solubility in two immiscible liquids, usually water and an organic solvent

**Longitudinal diffusion:** The diffusion of an analyte in the mobile phase as it passes through the analytical column driven by the concentration gradient. It contributes to band broadening, especially at low flow rates

**Matrix:** The components within a mixture that provide support and structure but are not directly relevant to the analytes of interest. Blood is an example of a matrix in the examination of drugs of abuse

**Mobile phase:** The phase that carries the analyte through the stationary phase and is used to influence the chromatographic separation

**Mode of separation:** Denotes the mechanism by which the separation takes place. It is characterised by the stationary phase and the solvents used to elute the analytes of interest. It can be classed as reversed phase, normal phase, ion exchange, and chiral chromatography

**Molecule:** The smallest part of a substance that is composed of two or more atoms of the same or different type held together by chemical forces

**Nernst distribution law (partition law):** The ratio (constant) at which an analyte will become distributed between two immiscible solvents at a defined temperature

**Normal phase chromatography:** The chromatographic separation in which the stationary phase is more polar than the mobile phase

**Orbital:** An electron cloud associated with an atom/molecule within which there is approximately a 95% probability of finding the associated electrons

**Partial charge:** A charge with a value of less than one measured unit (i.e., an electron). This is created due to the differences in electronegativity between the atoms forming the bond in a polar molecule

**Partition chromatography:** A distribution process in which a solute forms a homogeneous solution in each of the two phases (e.g., in paper chromatography the solute will be partitioned between the paper + water and the solvent)

**Pauli exclusion principle:** A maximum number of two electrons may be placed in each orbital and only when their spins are paired

**Peak area:** A measure of the area under the curve or peak within a chromatogram

**pKa:** Quantitative measure of the strength of an acid in solution. The larger the pKa value is, the smaller the extent of dissociation and therefore the weaker the acid is. A pKa < 2 = strong acid; pKa > 2 but < 7 = weak acid; pKa > 7 but < 10 = weak base; pKa > 10 = strong base

**Polarity:** The distribution of the electrical charge over the atoms that are joined together by the bond. In polar compound, the charge is distributed asymmetrically due to the differences in electronegativity between the atoms that make up the compound

**Precision:** The closeness of agreement between a series of measurements obtained from multiple sampling of the same sample. It can be considered at three levels: repeatability, intermediate precision, and reproducibility

**Pump:** In HPLC the pump drives the mobile phase through the system at a given, constant flow rate in the proportions determined by the user

**Qualitative:** An analysis in which identification of the analyte of interest is determined. This is usually achieved using a particular characteristic of the compound of interest, such as retention time, detector response (e.g., UV/Vis/fluorescence), and reference standard comparison

**Quality assurance:** The process of establishing whether a process or product meets customer expectations and is suitable for its intended purpose

**Quality control:** The systems that are put in place in order to ensure that the product is fit for its intended purpose

**Quantitative:** An analysis in which the amount of the analyte of interest is determined using a reference standard material of the same chemical structure

**Quantum theory:** The study of interactions of matter and radiation at the atomic and subatomic levels

**Range:** The range of an analytical method refers to the interval between the upper and lower concentration for which it has been demonstrated that there is a suitable level of accuracy, precision, and linearity

**Repeatability:** A measure of the precision of the method over a short period of time using the same sample solution

**Resistance to mass transfer:** The time taken for the analyte to transfer from the mobile to the stationary phase

**Resolution:** A measure of the separation between two adjacent compounds within a chromatographic separation. Under ideal conditions, resolution should be $\geq 1$ and $\leq 10$

**Retention factor:** A measure of the amount of time an analyte spends in the stationary phase relative to the mobile phase

**Retention time:** The time taken for an analyte to travel from the point of injection to the point of detection within an HPLC system

**Reverse phase chromatography:** Describes the chromatographic separation in which the stationary phase is nonpolar and the mobile phase is composed of an aqueous, moderately polar liquid

**Robustness:** A measure of a method's ability to withstand small but deliberate changes in the method parameters; it provides an indication of its reliability during normal usage

**Selectivity factor:** See *separation factor*

**Separation factor:** A measure of the ability of the system to separate two components within a mixture

**Solid phase extraction:** A process used to separate compounds from a mixture based on their chemical and physical characteristics

**Solubility:** A measure of the amount of solid required to be added to a given volume of solvent in order to form a saturated solution

**Solvent:** A substance that can dissolve another substance to form a solution. Water is an example of a commonly used solvent

**Specificity:** The ability to measure, without doubt, an analyte in the presence of other materials that might be expected to be present in the sample matrix

**Stationary phase:** The solid packing material contained within the column housing over which the mobile phase continuously flows. Choice of stationary phase influences the chromatographic separation

**Strong acid:** An acid that completely dissociates when dissolved in aqueous solution

**Strong base:**  A base that hydrolyses completely in aqueous solution, raising the pH toward 14

**Theoretical plate:**  A hypothetical zone within an HPLC column where the distribution between the two phases (stationary and mobile) takes place and is a measure of the effectiveness of the separating process. The greater the number of theoretical plates within a column is, the better the separating power

**Valence electrons:**  Electrons found in the outer of valance shell of an atom that determine the properties of the atom

**Valence shell:**  The outer shell of any atom. Electrons in the valence shell determine the chemical properties of the atom

**Valency:**  A measure of the number of chemical bonds that can be formed by the atoms within any given element

**Validation:**  In HPLC, this confirms that the method and the equipment consistently meet the requirements for a specific use and are fit for purpose

**Van der Waals forces:**  The weak electric forces of attraction or repulsion that exist between neutral molecules

**Weak acid/base:**  A substance that is only partially ionised in solution

# Index